British Railways Class 9F 2-10-0s

Compiled by Alan C. Butcher

Images from the Transport Treasury Archive

The · Transport · Treasury

Reviving the memories of yesterday…

© Images and design: The Transport Treasury 2023. Text Compiled by Alan C. Butcher

ISBN 978-1-913251-41-3

First published in 2023 by Transport Treasury Publishing Ltd., 16 Highworth Close, High Wycombe, HP13 7PJ

www.ttpublishing.co.uk

Printed in Malta by Gutenberg Press Ltd., Tarxien, GXQ 2902

Front cover: No 92044 heads a class C freight service over the troughs at Langley on 20th April 1960. The locomotive head code denotes a train of 'parcels, fish, livestock, milk, fruit or perishables' and made up of express passenger-rated wagons – in this case four-wheel box vans. Unlike a lot of freight vehicles at this time all the wagons would be braked enabling the train to run at faster speeds to ensure a timely arrival of short-life foodstuffs. No 92044 was delivered new to March depot on 12th January 1955 and withdrawn from Langwith Junction in April 1965.

Frontispiece: The former Crosti locomotives were classified in the power class of 8F due to their smaller diameter boilers, however this made little difference to their usage as No 92023 displays here as it climbs Ais Gill with a fully loaded anhydrite hopper train in tow during August 1964 whilst it was allocated to Carlisle Kingmoor depot. As with many of the class it ended its days at Birkenhead, being withdrawn in November 1967. *J. G. Walmsley*

Rear cover: No 92015 heads a class H mixed freight southwards along the Midland main line to the north of St Albans on 23rd September 1955 whilst the locomotive was allocated to Wellingborough. No 92015 ended its days at Carlisle Kingmoor depot in April 1967. *(1171)*

Right: Standing outside No 10 Shop at Crewe Works, No 92037 is destined for the paint shop on 28th November 1954. The locomotive would enter service on the 8th December, being allocated to New England shed at Peterborough where it would remain until June 1963 when it was moved to Immingham. It would be placed into store in January 1965, being withdrawn the following month and consigned to Albert Draper's scrapyard at Hull. Cut up on 10th June, it was one of 37 to be recycled at the yard. *(1422)*

Contents

Introduction

It could all have been so different! The initial proposals for a BR Standard freight locomotive were of a 2-8-2 wheel arrangement, with designs being produced in 1948 (with bar frames, and a number of parts being interchangeable with a proposed Class 5 mixed-traffic Pacific) and again in 1951 (with plate frames and parts interchangeable with the 'Britannia' Pacifics) with 5ft 3in diameter driving wheels. Later in 1951 a Class 8 2-10-0 followed this variation, incorporating 5ft diameter wheels with roller bearings on all axles. Interestingly, all the diagrams for these proposals incorporated the 'Austerity'-style chimney as late as 1953. The 'definitive' 9F-variation appeared late in 1953 incorporating dome regulator and plain bearings.

The actual design was allocated to Brighton Works, although none of the class was to be built there. Of the total of 251, 198 were built at Crewe and the remainder at Swindon including the last one of all, No 92220, that was named *Evening Star* in a ceremony at the works on 18th March 1960. It was the 999th BR Standard, and last steam locomotive, built for use on British Railways.

Correspondence between the Western Region and Railway Executive in 1953 states that the case for the construction of the 9Fs was based on a 45-year life span; little were those involved to know that within 10 years the steam era was virtually over. It is interesting to note that the initial thoughts for the run-down of steam anticipated some 5,000 locomotives remaining in service in 1975 with final elimination in 1985, although the latter date could stretch until 1990.

The only major difference in the appearance of the class were the 10 locomotives incorporating the ideas of the Italians Atillio Franco and Peiro Crosti. Many Italian locomotives were fitted with 'pre-heaters' – basically a secondary multi-tube boiler to improve efficiency. Design work was carried out at Brighton Works, with construction at Crewe. See the section of 'Crime of the Crostis' on pages 24-28. Due to the length of time some of the Crosti-variant were in store, the only year that the entire class was in service was 1963; the first of the standard 9Fs were withdrawn in May 1964 – the last in June 1968.

Construction

The construction of the 9Fs was split between Crewe and Swindon as follows:

92000-14, Crewe, order No E487 – to traffic 01/54-05/54
92015-19, Crewe, order No E491 – to traffic 09/54-10/54
92020-29, Crewe, order No E488 – to traffic 03/55-07/55*
92030-49, Crewe, order No E489 – to traffic 11/54-03/55
92050-86, Crewe, order No E490 – to traffic 08/55-06/56
92087-96, Swindon, Lot No 421 – to traffic 08/56-04/57
92097-134, Crewe, order No E493 – to traffic 06/56-05/57
92135-177, Crewe, order No E494 – to traffic 06/57-03/58
92178-202, Swindon, Lot No 422 – to traffic 09/57-12/58
92203-220, Swindon, Lot No 429 – to traffic 04/59-03/60
92221-250, Crewe, order No E497 – to traffic 05/58-12/58
*Crosti variant

The split of construction between Crewe and Swindon shows that the average cost of the locomotives ordered in 1954 were Crewe, £22,500, against Swindon's £30,200. The authorised cost was £23,500 per locomotive. Interestingly all cylinders were cast at Crewe, along with all the pressed boilerplates.

Tenders

As built these were Region specific, with the variations given below as ex-works:

BR1B – LMR (Crosti) / NER*
BR1C – LMR (standard)
BR1G – WR
BR1F – ER
BR1K – LMR (mechanical stoker)
*Not fitted with water pick-up apparatus

It was planned that all tenders would be constructed at the same works as the locomotives, however it was reported that some of the final batch of Swindon-built examples were manufactured at Ashford Works.

Tender type (No built for 9Fs)	Tender style	Water capacity	Coal capacity	Weight in working order
BR1B (20)	Flush	4,725	7 tons	51.25 tons*
BR1C (85)	Flush	4,725	9 tons	53.25 tons*
BR1F (85)	Flush	5,625	7 tons	55.25 tons
BR1G (58)	Inset at top	5,000	7 tons	52.50 tons
BR1K (3)	Flush	4,325	9 tons	52.35 tons

*Visibly identical, the coal capacity could – in theory – be changed by moving the coal partition, being nearer the cab in the case of the BR1B.

As with most classes tender types were sometimes changed so photographic references are useful to the modeller; for example No 92090 ran with BR1F, BR1G and BR1C-designs during its career operating from NER and LMR depots.

As a photographic record this is not the place to give a complete history of the class, and those wishing for additional details are directed to the RCTS book on the class – *A Detailed History of British Railways Standard Steam Locomotives: Volume 4: 9F 2-10-0s*.

Note on dates: The BR practice was to use the 'period ending' dates, using the Saturday as the end date. So unless the actual date is quoted then the dates recorded are either week ending, four week ending; or in some cases five or six weeks ending. Rather than scatter the text with numerous 'week ending', a looser description is given and readers are referred to the data tables and the RCTS book mentioned above.

Bibliography and Further Reading

Atkins, P., *The British Railways Standard 9F 2-10-0*, Irwell Press, 9781871608298, 1993.

Bond, Roland C., *A Lifetime with Locomotives*, Goose & Son Publishers, 0900404302, 1975.

Cornwell, E. L., (Ed), *Locomotives Illustrated: 5 – 9F 2-10-0s*, Ian Allan, 1976.

Cox, E. S., *British Railways Standard Steam Locomotives*, Ian Allan, no ISBN, 1966.

Griffiths, R. & Smith, P., *The Directory of British Engine Sheds and Principal Locomotive Servicing Points: Vol 1*, 9780860935421, OPC, 1999.

Griffiths, R. & Smith, P., *The Directory of British Engine Sheds and Principal Locomotive Servicing Points: Vol 2*, 9780860935483, OPC, 2000.

Grindlay, Jim, *British Railways Steam Locomotive Allocations 1948-1968: Part 5 BR Standard & Ex-War Department 70000-92250*, Modelmaster Publications, 9780954426253, 2006.

Haresnape, B., *Ivatt & Riddles Locomotives: A Pictorial History*, Ian Allan, 0711007950, 1977.

Longworth, H., *British Railways Steam Locomotives 1948-1968,* 9780860935933, OPC, 2005.

Longworth, H., *British Railways Steam Locomotive Allocations*, 9780860936428, OPC, 2011.

RCTS, *A Detailed History of British Railways Standard Steam Locomotives: Vol 1 – Background to Standardisation and the Pacific Classes*, RCTS, 0901115819, 1994.

RCTS, *A Detailed History of British Railways Standard Steam Locomotives: Vol 4 – 9F 2-10-0s*, RCTS, 0901115959, 2008.

RCTS, *A Detailed History of British Railways Standard Steam Locomotives: Vol 5 – The End of an Era*, RCTS, 0901115973, 2012.

Rogers, H. C. B., *Riddles and the 9Fs*, 9780711012080, Ian Allan, 1982.

Scotney, I. & Egan, B., *British Railways Locomotives cut up by Drapers of Hull*, Hutton Press, 0907033830, 1989.

Scott Morgan, J., *The '9Fs': BR's Heavy Freight Locomotives*, 9780711022652, Ian Allan, 1994.

Walmsley, T., *Shed by Shed, Part One, London Midland*, 9780956061553, 2010.

Walmsley, T., *Shed by Shed, Part Two, Eastern*, 9780956061560, 2010.

Walmsley, T., *Shed by Shed, Part Three*, North Eastern, 9780956061539, 2010.

Walmsley, T., *Shed by Shed, Part Four*, Scottish, 9780956061577, 2011.

Walmsley, T., *Shed by Shed, Part Six*, Western, 9780956061522, 2009.

Williams, A., *BR Standard Steam Album*, Ian Allan, 0711010102, 1980.

–, *abc British Railways Locomotives: Combined Volume, Winter 1955/56*, Ian Allan, 9780711005068, reprinted 1999.

–, *abc British Railways Locomotives: Combined Volume, Summer 1957*, Ian Allan, 9780711038455, reprinted 2016.

–, *abc British Railways Locomotives: Combined Volume, Summer 1958*, Ian Allan, 9780711037694, reprinted 2013.

Amongst many others, the following websites were useful during preparation of this title:
www.brdatabase.info
www.gracesguide.co.uk

A member of the shed staff appears to have caught the photographer in the act at Toton depot on 27th April 1958. New to traffic on 11th October 1955, No 92057 was built at Crewe and allocated to Toton where it remained until January 1960 when it moved to Westhouses. It was one of only a handful of the class to be overhauled at Gorton Works, undergoing a Classified Light there between 9th January and 11th February 1962 whilst allocated to Annersley. Its final year was spent at Birkenhead, having arrived in April 1965 it was withdrawn week ending 2nd October 1965. Following a short period in storage it was sold to T. W. Ward at Beighton, Sheffield, and demolished in December the same year. Details of No 92160 can be found on page 72.

Data Panel

Ann	Annersley		GHR	Gloucester Horton Road		Salt	Saltley			
Ban	Banbury		Imm	Immingham		Shef	Sheffield Darnall			
Bath	Bath Green Park		Ket	Kettering		Sout	Southall			
BBR	Bristol Barrow Road		KiA	Kirkby-in-Ashfield		SpJ	Speke Junction			
BMS	Birkenhead Mollington Street		LaJ	Langwith Junction		SPM	St Philip's Marsh			
CarC	Cardiff Canton		Leic	Leicester		STJ	Severn Tunnel Junction			
CaK	Carlisle Kingmoor		Mar	March		Tot	Toton			
Car	Carnforth		Mex	Mexborough		TyD	Tyne Dock			
CED	Cardiff East Dock		NeH	Newton Heath		WaD	Warrington Dallam			
Col	Colwick		NEJ	Newport Ebbw Junction		Wel	Wellingborough			
Cri	Cricklewood		New	New England		Wes	Westhouses			
Der	Derby		Nor	Northampton		York	York (North)			
Don	Doncaster		OOC	Old Oak Common		Tys	Tyseley			
Frod	Frodingham		PonR	Pontypool Road		Wak	Wakefield			

Number	To Traffic on	Original Tender Type	First Shed Allocation	Final Shed Allocation	Withdrawn (*on)	Number	To Traffic on	Original Tender Type	First Shed Allocation	Final Shed Allocation	Withdrawn (*on)
92000	05/01/1954	1G	NEJ	GHR	*02/07/1965	92039	17/12/1954	1F	New	LaJ	17/10/1965
92001	12/01/1954	1G	NEJ	Wak	*19/01/1957	92040	23/12/1954	1F	New	LaJ	08/08/1965
92002	19/01/1954	1G	NEJ	BMS	11/11/1967	92041	31/12/1954	1F	New	LaJ	29/08/1965
92003	31/01/1954	1G	NEJ	CED	*26/03/1965	92042	21/01/1955	1F	New	Col	*19/12/1965
92004	30/01/1954	1G	NEJ	Car	23/03/1968	92043	18/01/1955	1F	Mar	CaK	30/07/1966
92005	05/02/1954	1G	NEJ	York	30/08/1965	92044	29/01/1955	1F	Mar	LaJ	11/04/1965
92006	10/02/1954	1G	NEJ	Wak	10/04/1967	92045	09/02/1955	1C	Wel	BMS	16/09/1967
92007	28/02/1954	1G	NEJ	GHR	*31/12/1965	92046	11/02/1955	1C	Wel	BMS	21/10/1967
92008	31/03/1954	1G	Wel	WaD	07/10/1967	92047	18/02/1955	1C	Wel	BMS	11/11/1967
92009	31/03/1954	1G	Wel	Car	23/03/1968	92048	26/02/1955	1C	Wel	BMS	16/09/1967
92010	31/05/1954	1F	Mar	CaK	02/04/1966	92049	10/03/1955	1C	Wel	BMS	11/11/0967
92011	31/05/1954	1F	Mar	BMS	11/11/1967	92050	19/08/1955	1C	Tot	Wad	30/09/1967
92012	31/05/1954	1F	Mar	CaK	28/10/1967	92051	22/08/1955	1C	Tot	CaK	22/10/1967
92013	31/05/1954	1F	Mar	Salt	01/10/1966	92052	26/08/1955	1C	Tot	CaK	12/08/1967
92014	31/05/1954	1F	Mar	BMS	14/10/1967	92053	17/09/1955	1C	Tot	Wad	19/02/1966
92015	30/09/1954	1C	Wel	CaK	29/04/1967	92054	21/09/1955	1C	Tot	SpJ	04/05/1968
92016	31/10/1954	1C	Wel	Car	21/10/1967	92055	22/09/1955	1C	Tot	SpJ	23/12/1967
92017	31/10/1954	1C	Wel	CaK	23/12/1967	92056	14/10/1955	1C	Tot	CaK	11/11/1967
92018	31/10/1954	1C	Wel	CaK	22/04/1967	92057	11/10/1955	1C	Tot	BMS	02/10/1965
92019	31/10/1954	1C	Wel	CaK	10/06/1967	92058	13/10/1955	1C	Tot	CaK	04/11/1967
92020	21/03/1955	1B	Wel	BMS	21/10/1967	92059	20/10/1955	1C	Tot	BMS	17/06/1966
92021	21/03/1955	1B	Wel	BMS	11/11/1967	92060	05/11/1955	1B	TyD	TyD	*23/10/1966
92022	21/03/1955	1B	Wel	BMS	11/11/1967	92061	12/11/1955	1B	TyD	TyD	04/09/1966
92023	21/03/1955	1B	Wel	BMS	11/11/1967	92062	19/11/1955	1B	TyD	TyD	26/06/1966
92024	11/06/1955	1B	Wel	BMS	11/11/1967	92063	25/11/1955	1B	TyD	TyD	24/11/1966
92025	18/06/1955	1B	Wel	BMS	11/11/1967	92064	02/12/1955	1B	TyD	TyD	24/11/1966
92026	18/06/1955	1B	Wel	BMS	11/11/1967	92065	10/12/1955	1B	TyD	WaK	03/04/1967
92027	25/06/1955	1B	Wel	SpJ	05/08/1967	92066	15/12/1955	1B	TyD	TyD	*23/05/1965
92028	09/07/1955	1B	Wel	Salt	22/10/1966	92067	20/12/1955	1F	Don	Ban	05/11/1966
92029	09/07/1955	1B	Wel	BMS	11/11/1967	92068	29/12/1955	1F	Don	Der	15/01/1966
92030	07/11/1954	1F	New	Wak	*27/02/1967	92069	31/12/1955	1F	Don	SpJ	05/05/1966
92031	11/11/1954	1F	Mar	NeH	07/01/1967	92070	14/01/1956	1F	Don	BMS	11/11/1967
92032	17/11/1954	1F	New	BMS	15/04/1967	92071	21/01/1956	1F	Don	CaK	11/11/1967
92033	19/11/1954	1F	New	Nor	18/09/1965	92072	01/02/1956	1F	Don	KiA	08/01/1966
92034	02/12/1954	1F	New	Imm	*31/05/1964	92073	09/02/1956	1F	Don	BMS	11/11/1967
92035	03/12/1954	1F	New	Imm	*10/02/1966	92074	18/02/1956	1F	Don	CaK	15/04/1967
92036	02/12/1954	1F	New	New	*06/12/1964	92075	01/03/1956	1F	Don	CaK	17/09/1967
92037	08/12/1954	1F	New	Imm	21/02/1965	92076	06/03/1956	1F	Don	CaK	25/02/1967
92038	14/12/1954	1F	New	LaJ	11/04/1965	92077	13/03/1956	1C	Tot	Car	22/06/1968

Number	To Traffic on	Original Tender Type	First Shed Allocation	Final Shed Allocation	Withdrawn (*on)	Number	To Traffic on	Original Tender Type	First Shed Allocation	Final Shed Allocation	Withdrawn (*on)
2078	21/03/1956	1C	Tot	Wad	06/05/1967	92139	28/06/1957	1C	Salt	CaK	09/09/1967
2079	29/03/1956	1C	Tot	BMS	11/11/1967	92140	05/07/1957	1F	New	LaJ	11/04/1965
2080	14/04/1956	1C	Tot	CaK	06/05/1967	92141	29/07/1957	1F	New	Col	*19/12/1965
2081	23/04/1956	1C	Tot	NeH	12/02/1966	92142	29/07/1957	1F	New	New	21/02/1965
2082	03/05/1956	1C	Wel	BMS	11/11/1967	92143	10/08/1957	1F	New	New	21/02/1965
2083	11/05/1956	1C	Wel	BMS	18/02/1967	92144	19/08/1957	1F	New	Col	*19/12/1965
2084	19/05/1956	1C	Wel	BMS	11/11/1967	92145	26/08/1957	1F	New	Imm	*06/02/1966
2085	26/05/1956	1C	Wel	BMS	24/12/1966	92146	05/09/1957	1F	New	Don	17/04/1966
2086	06/06/1956	1C	Wel	BMS	11/11/1967	92147	14/09/1957	1F	New	Imm	04/04/1965
2087	29/08/1956	1F	Don	Car	25/02/1967	92148	21/09/1957	1F	New	Col	*19/12/1965
2088	24/10/1956	1F	Don	Car	27/04/1968	92149	01/10/1957	1F	New	LaJ	27/06/1965
2089	26/09/1956	1F	Don	BMS	18/02/1967	92150	05/10/1957	1C	Wes	Wak	*17/04/0967
2090	08/11/1956	1F	Don	BMS	20/05/1967	92151	09/10/1957	1C	Salt	BMS	22/04/1967
2091	27/11/1956	1F	Don	Car	25/05/1968	92152	16/10/1957	1C	Salt	BMS	11/11/1967
2092	13/12/1956	1F	Don	BMS	22/10/1966	92153	21/10/1957	1C	Tot	SpJ	21/01/1968
2093	15/01/1957	1F	Don	CaK	02/09/1967	92154	28/10/1957	1C	Wel	SpJ	22/07/1967
2094	04/02/1957	1F	Don	SpJ	04/05/1968	92155	11/05/1957	1C	Salt	SpJ	19/11/1966
2095	06/03/1957	1F	Ann	Wad	01/10/1966	92156	15/11/1957	1C	Tot	Wad	22/07/1967
2096	02/04/1957	1F	Ann	CaK	25/02/1967	92157	20/11/1957	1C	Tot	BMS	19/08/1967
2097	27/06/1956	1B	TyD	TyD	23/10/1966	92158	22/11/1957	1C	Tot	SpJ	23/07/1966
2098	04/07/1956	1B	TyD	TyD	31/07/1966	92159	27/11/1957	1C	Wel	BMS	22/07/1967
2099	23/07/1956	1B	TyD	TyD	04/09/1966	92160	29/11/1957	1C	Wel	Car	29/06/1968
2100	30/07/1956	1C	Tot	BMS	13/05/1967	92161	17/12/1957	1C	Wes	CaK	10/12/1966
2101	13/08/1956	1C	Tot	BMS	14/10/1967	92162	19/12/1957	1C	Wes	BMS	11/11/1967
2102	15/08/1956	1C	Tot	BMS	11/11/1967	92163	24/03/1958	1C	Ket	BMS	11/11/1967
2103	18/08/1956	1C	Tot	BMS	27/05/1967	92164	01/04/1958	1C	Leic	Salt	30/07/1966
2104	24/08/1956	1C	Tot	BMS	25/02/1967	92165	17/04/1958	1K	Salt	SpJ	16/03/1968
2105	07/09/1956	1C	Wel	BMS	14/01/1967	92166	24/05/1958	1K	Salt	BMS	11/11/1967
2106	12/09/1956	1C	Wel	BMS	29/07/1967	92167	09/05/1958	1K	Salt	Car	29/06/1968
2107	20/09/1956	1C	Wel	BMS	25/02/1967	92168	20/12/1957	1F	Don	Don	27/06/1965
2108	12/10/1956	1C	Wel	BMS	11/11/1967	92169	27/12/1957	1F	Don	Don	31/05/1964
2109	15/10/1956	1C	Tot	BMS	11/11/1967	92170	31/12/1957	1F	Don	Don	31/05/1964
2110	23/10/1956	1C	Tot	CaK	31/12/1967	92171	01/02/1958	1F	Don	New	31/05/1964
2111	06/11/1956	1C	Cri	BMS	28/10/1967	92172	29/01/1958	1F	Don	Don	17/04/1966
2112	17/11/1956	1C	Cri	BMS	11/11/1967	92173	10/02/1958	1F	Don	Don	06/03/1966
2113	24/11/1956	1C	Wes	BMS	07/10/1967	92174	10/02/1958	1F	Don	Don	12/12/1965
2114	29/11/1956	1C	Wes	CaK	22/07/1967	92175	21/02/1958	1F	Don	Don	31/05/1964
2115	12/12/1956	1C	Wes	SpJ	19/02/1966	92176	04/03/1958	1F	Don	New	31/05/1964
2116	22/12/1956	1C	Wes	Wad	12/11/1966	92177	12/03/1958	1F	Don	Don	31/05/1964
2117	27/12/1956	1C	Wes	SpJ	23/12/1967	92178	28/09/1957	1F	New	LaJ	*03/10/1965
2118	28/12/1956	1C	Wes	Car	25/05/1968	92179	07/10/1957	1F	New	Col	*14/11/1965
2119	12/01/1957	1C	Wes	CaK	23/09/1967	92180	08/11/1957	1F	New	LaJ	11/04/1965
2120	28/01/1957	1C	Wes	BMS	08/07/1967	92181	22/11/1957	1F	New	New	21/02/1965
2121	08/02/1957	1C	Wel	BMS	29/07/1967	92182	13/12/1957	1F	New	Don	17/04/1966
2122	18/02/1957	1C	Wel	BMS	11/11/1967	92183	20/12/1958	1F	New	Don	03/04/1966
2123	26/02/1957	1C	Wel	BMS	28/10/1967	92184	10/01/1958	1F	New	Imm	*21/02/1965
2124	02/03/1957	1C	Wel	Wad	03/12/1966	92185	05/01/1958	1F	New	Imm	21/02/1965
2125	09/03/1957	1C	Wel	CaK	30/12/1967	92186	27/01/1958	1F	New	LaJ	29/08/1965
2126	18/03/1957	1C	Wel	Wad	05/08/1967	92187	12/02/1958	1F	New	Col	21/02/1965
2127	29/03/1957	1C	Wel	BMS	19/08/1967	92188	27/02/1958	1F	New	Col	21/02/1965
2128	05/04/1957	1C	Tot	Car	11/11/1967	92189	14/03/1958	1F	Mex	Col	*19/12/1965
2129	01/04/1957	1C	Salt	CaK	01/07/1967	92190	28/03/1958	1F	Mex	Don	10/10/1965
2130	15/04/1957	1C	Salt	CaK	14/06/1966	92191	15/04/1958	1F	Shef	Col	*19/12/1965
2131	25/04/1957	1C	Salt	BMS	23/09/1967	92192	01/05/1958	1F	Don	Frod	21/02/1965
2132	29/04/1957	1C	Salt	CaK	21/10/1967	92193	23/05/1958	1F	Don	Imm	13/06/1965
2133	22/05/1957	1C	Salt	BMS	22/07/1967	92194	10/06/1958	1F	Don	Imm	05/12/1965
2134	24/05/1957	1C	Salt	BMS	10/12/1966	92195	27/06/1958	1F	Don	LaJ	09/05/1965
2135	11/06/1957	1C	Salt	WaK	*08/06/1967	92196	15/08/1958	1F	Don	Imm	06/12/1964
2136	18/06/1957	1C	Salt	Salt	29/10/1966	92197	11/09/1958	1F	Don	Imm	19/09/1965
2137	24/06/1957	1C	Salt	CaK	09/09/1967	92198	06/10/1958	1F	Don	Frod	02/08/1964
2138	25/06/1957	1C	Salt	SpJ	08/07/1967	92199	29/10/1958	1F	Don	Frod	02/08/1964

Number	To Traffic on	Original Tender Type	First Shed Allocation	Final Shed Allocation	Withdrawn (*on)	Number	To Traffic on	Original Tender Type	First Shed Allocation	Final Shed Allocation	Withdrawn (*on)
92200	18/11/1958	1F	Don	LaJ	03/10/1965	92226	26/06/1958	1G	Ban	STJ	*10/09/1965
92201	05/12/1958	1F	Don	Don	27/03/1966	92227	03/07/1958	1G	Ban	SpJ	04/11/1967
92202	27/12/1958	1F	Don	Imm	19/12/1965	92228	12/07/1958	1G	Ban	SpJ	28/02/1967
92203	06/04/1959	1G	SPM	BMS	11/11/1967	92229	18/07/1958	1G	Ban	NEJ	*06/11/1964
92204	21/04/1959	1G	SPM	SpJ	09/12/1967	92230	01/08/1958	1G	Ban	GHR	*31/12/1965
92205	01/05/1959	1G	SPM	Wak	08/06/1967	92231	01/08/1958	1G	PonR	York	17/11/1966
92206	06/05/1959	1G	SPM	Wak	09/05/1967	92232	06/08/1958	1G	PonR	CED	*26/12/1964
92207	01/06/1959	1G	SPM	NEJ	*23/12/1964	92233	11/08/1958	1G	PonR	SpJ	03/02/1968
92208	11/06/1959	1G	Laira	CaK	28/11/1967	92234	19/08/1958	1G	PonR	BMS	11/11/1967
92209	26/06/1959	1G	Laira	Bath	*31/12/1965	92235	22/08/1958	1G	PonR	BBR	*18/11/1965
92210	07/08/1959	1G	CarC	NEJ	*06/11/1964	92236	04/09/1958	1G	PonR	STJ	*21/04/1965
92211	08/09/1959	1G	OOC	Wak	22/05/1967	92237	09/09/1958	1G	NEJ	NEJ	*10/09/1965
92212	22/09/1959	1G	Ban	Car	06/01/1968	92238	22/09/1958	1G	NEJ	STJ	*10/09/1965
92213	22/10/1959	1G	SPM	Ban	05/11/1966	92239	25/09/1958	1G	NEJ	York	*17/11/1966
92214	30/10/1959	1G	CarC	STJ	*09/08/1965	92240	01/10/1958	1G	NEJ	Sout	*06/08/1965
92215	17/11/1959	1G	Ban	Wak	*08/06/1967	92241	06/10/1958	1G	NEJ	Sout	*02/07/1965
92216	01/12/1959	1G	CarC	STJ	*26/10/1965	92242	13/10/1958	1G	NEJ	STJ	*21/05/1965
92217	22/12/1959	1G	CarC	Tys	30/07/1966	92243	21/10/1958	1G	NEJ	Bath	*01/01/1966
92218	08/01/1960	1G	SPM	SpJ	04/05/1968	92244	28/10/1958	1G	NEJ	GHR	*31/12/1965
92219	27/01/1960	1G	SPM	CED	*09/08/1965	92245	04/11/1958	1G	OOC	Sout	*26/12/1964
92220	25/03/1960	1G	CarC	CED	*26/03/1965	92246	20/11/1958	1G	OOC	GHR	*31/12/1965
92221	30/05/1958	1G	Ban	York	*10/05/1965	92247	02/12/1958	1G	OOC	NeH	08/10/1958
92222	02/06/1958	1G	Ban	Sout	*26/03/1965	92248	05/12/0958	1G	NEJ	CED	*21/05/1965
92223	06/06/1958	1G	Ban	Car	13/04/1968	92249	10/12/1958	1G	NEJ	SpJ	04/05/1968
92224	16/06/1958	1G	Ban	Wad	30/09/1967	92250	16/12/1958	1G	Ban	GHR	*31/12/1965
92225	20/06/1958	1G	Ban	NEJ	*02/07/1965						

All change at Ais Gill on 7th August 1965. An unidentified 9F stands outside the signal box as the locomotive crews change over. Following closure of the box it was preserved, and, having been relocated, can be seen at Butterley station on the heritage Midland Railway. *(1004)*

Chapter 1: The class of 1954: Nos 92000-19, 92030-41

On 7th August 1965, Tyseley-based No 92001 was 'borrowed' to work the 11.10am Bournemouth-Newcastle service via the Great Central main line. It is seen here at Basingstoke on the former London & South Western Railway line, the head code for the Bournemouth route service matches the express passenger code used on the other regions so no change would be necessary for its journey over the Great Western line to Reading. To traffic on 12th January 1954, it was built with the then standard single chimney, receiving the double version during a Heavy Intermediate overhaul at Swindon between 29th September 1959 and 10th March 1960. Withdrawn on 17th January 1967, it was the only 9F to be recycled by Cox & Danks at Wadsley Bridge the following May. *(1013)*

Summer 1966 and No 92002 stands alongside the shed building at Crewe South in the company of a fellow class member and an Ivatt Class 2MT Mogul on the 18th August. No 92002 is missing its Tyseley shed code and self-cleaning smokebox plates. Allocated to Newport Ebbw Junction when new in January 1954, the locomotive first arrived at Tyseley depot in May 1963, returning via Banbury depot in November 1964. It arrived at its final depot of Birkenhead in December 1966, surviving ....rvice until November 1967 when the bulk of the class remaining in service were withdrawn. No 92002 was sold to ....bell, Airdrie, for scrapping in March 1968 and was cut up the following month. *James L. Stevenson*

New to traffic on 30th January 1954, and allocated to Newport Ebbw Junction depot, No 92004, after several other moves, arrived at Banbury shed in March 1963. It is seen here as it heads south towards Banbury in June 1966 at the head of a box van, three mineral wagons and a brake van. By this date the cast smokebox number plate had been replaced with white paint. The locomotive remained at Banbury until moved to Carlisle Kingmoor in October 1966. Withdrawal would occur from Carnforth depot in March 1968, being recycled at Newport by John Cashmore's cutters five months later. *(1033)*

Looking like a victim of a 'rough shunt', No 92005 stands in Swindon Works on 18th August 1957, its coupling rods having been removed for the journey. At this date it was allocated to Newport Ebbw Junction where it had arrived newly built from Crewe in February 1954. Its original single blast pipe arrangement would be replaced with a double one during a visit to Swindon for a Heavy Intermediate overhaul between 15th December 1959 and 20th May 1960. An early withdrawal, on 30th August 1965 at York depot, it was scrapped in Sheffield by T. W. Ward at Beighton that December. *(1047)*

Left: It was unusual for photographs to be taken within the confines of Kingmoor depot, however, No 92006 is seen on the turntable inside the roundhouse on August 1964, sporting a double chimney that had been fitted during a Heavy General overhaul at Swindon between 23rd February and 21st September 1960. In September 1964 it would enter Crewe Works for an Unclassified overhaul that lasted until 23rd January 1961, a further Intermediate overhaul later the same year would see the locomotive through until withdrawal in April 1967.
John G. Walmsley (1055)

Below: With a combined power rating of 17F, No 92008 and Stanier Class 8F No 48477 stand side by side within the confines of Willesden depot on 20th June 1954 – the 9F had worked in from its home shed of Wellingborough whilst the 8F was allocated to Bescot. The 9F underwent its one and only Heavy General overhaul at Crewe between 11th October and 5th November 1960 and following four Light Casual repairs would end its days at Warrington, being withdrawn in October 1967 and scrapped in South Wales in August 1968.
Arthur Carpenter

Above: Water Orton is the location of No 92009 as it heads eastwards through the overbridge at the east end of the station. It arrived at Saltley depot from Burton in March 1958 and, apart from a short loan period to Annesley, stayed until November 1959 when it moved to Rowsley. Over the years it was one of 55 members of the class allocated to Saltley. The first station that served Water Orton was opened in 1842 by the Birmingham & Derby Junction Railway on its line into Birmingham Lawley Street from Whitacre Junction. However, the Midland Railway built a cut-off line from slightly further west to a junction at Kingsbury between 1908 and 1909; the station was resited in August 1908. Although the distance saved was only a mile-and-a-quarter, the junctions at Water Orton and Kingsbury could be taken at a much higher speed than the original one at Whitacre. *(1080)*

Opposite Top: No 92010 awaits departure time at Cambridge on 8th September 1954 with the 9.45am Temple Mills to Whitemoor service. The 9F arrived ex-works at March in May 1954, remaining until January 1957 when it moved to Annesley. The marshalling yard at Whitemoor was opened by the LNER on 3rd March 1929. It had 43 sorting sidings, providing room for 4,000 wagons for 350 destinations. In 1933 a down yard was added and, by the start of World War 2, 8,000 wagons could be sorted each day. In the 1960s line and station closures meant that the need for these large marshalling yards was no longer needed, so on 27th January 1972 the down yard closed with the up yard taking over its duties. By 1982 the Departure sidings were now only used and what little traffic there still was transferred to Peterborough. *(1105)*

Opposite Bottom: No 92011 heads a rake of mineral wagons southbound through Nottingham Victoria. The locomotive was allocated to Annesley depot from May 1957 to August 1965 when it moved to Birkenhead. Withdrawal came from there in November 1967. The Great Central and Great Northern railways shared Nottingham Victoria station. Interestingly the joint owners failed to reach an agreement on what their station was to be named. The Town Clerk tried to resolve the situation by suggesting the name 'Nottingham Victoria' to reflect the fact that the planned opening date coincided with Queen Victoria's birthday. The last through service from Nottingham to London ran on 3rd September 1966. All that was left was a DMU service between Nottingham and Rugby. Victoria station was finally closed on 4th September 1967 when this service was cut back to Arkwright Street to allow for the building of the Victoria Shopping Centre and flats, leaving only the clock tower surviving. *(1117)*

Above: 9F action at Southall as No 92012 is seen exiting the freight sidings and crossing the main lines as it does so in August 1961. The gasholders were part of Southall Gas Works originally constructed by the Brentford Gas Company, opening in 1869. It was required to meet rapidly increasing demand in Middlesex, which outstripped the capacity of the company's original works on the Thames at Brentford. The gas works was originally established at the western end of the full site, and progressively expanded to the east over sites originally used for brickyards and industrial works. Following nationalisation of the gas industry in 1949 the plant came under the control of the North Thames Gas Board. Construction of an oil gasification plant began and by 1951 up to 300,000cu ft (8,500m^3) of gas a day was being produced in this way, primarily at times of peak demand. In the early 1960s a Carbureted Water-Gas Plant was installed at Southall to make town gas from a feedstock of liquefied petroleum gas obtained from Fawley refinery via a 70-mile pipeline. With the move to North Sea gas the works closed in 1973, leaving gas distribution and storage as the main on site functions. During the early 21st century, construction began on The Green Quarter. At the date of the image No 92012 was allocated to Annesley depot that was well known for its 'windcutter' service between there and Woodford Halse. It stayed at Annesley until moving to Rowsley in July 1964. Later moving to Carlisle Kingmoor via Kettering, it was withdrawn at the end of October 1967. *(1128)*

Opposite Top: No 92014 ex-works at Crewe in May 1954. It entered traffic on the 18th of that month and was allocated to the Eastern Region's depot at March, although it went via the International Railway Exhibition at Willesden where it was on display with No 71000 *Duke of Gloucester* and the prototype gas turbine locomotive No 18000; both the latter are now preserved. The exhibition included stock from London Underground, various pieces of permanent way and signalling equipment, including some of the UK's newest tech. At the time, containerised transport was in its infancy, having been pioneered by UK railways with wooden containers that were craned between rail wagons and road transport. The displays even included Wickham railcars for maintenance crews. Withdrawn in October 1967, No 92014 was scrapped in August the following year by J. Buttigieg at Newport. *(1165)*

Opposite Bottom: No 92016 drags a rake of open mineral wagons along the Midland main line near Elstree on 14th May 1955. The locomotive was allocated to Wellingborough at the time, having been delivered from Crewe Works on the 7th October the previous year. Just over 13 years later it was withdrawn from Carnforth depot having spent time at Saltley, Newton Heath and Bolton. It was sold to J. McWilliams of Shettleston for disposal in April 1968 and scrapped almost immediately on arrival at the yard. *(1178)*

Opposite Top: No 92017 runs south through Royton Junction station, near Oldham, with an excursion to Rhyl on 14th September 1958. The 9F was allocated to Newton Heath at the time, and the depot was not averse to using these freight locomotives on excursion traffic with a number being recorded in the late 1950s. As with a number of the class, No 92017 ended its days at Carlisle Kingmoor, withdrawal taking place in December 1967. Royton Junction station was on the Oldham Loop Line in Greater Manchester. It opened on 1st July 1864, and was the junction for the short branch line to Royton. The branch to Royton was closed to goods on 2nd November 1964, and to passengers on 16th April 1966. A 450-yard section of the line remained in use for freight traffic from Royton Junction to Higginshaw Gas Sidings, this final section closed on 6th April 1970. The main line station was renamed Royton on 8th May 1978, and was closed from 11th May 1987. Approval for closure had been given on 8th May 1987, as the station had been replaced by a new station at Derker that opened half-a-mile (0.8km) away on 30th August 1985. *(1184)*

Opposite Bottom: Coaled and ready for a day's work, No 92019 is seen on shed at Carlisle Kingmoor on 17th July 1965. The locomotive had been delivered new to Wellingborough in October 1954 and must have felt at home there, as during the course of its career was allocated there on four occasions. Its final move away came in January 1964 when it was moved to Kettering and then onto Kingmoor, its final depot, in June 1964. After three years at the depot it was withdrawn in June 1967 and was sold to the Motherwell Machinery & Scrap Co at Wishaw where it was cut up in January 1968. *(1199)*

Below: Allocated new to Peterborough's New England depot on delivery from Crewe Works in November 1954, No 92030 was transferred to Annesley in May 1957 and is seen carrying that depot's shed code plate on 24th May 1964 following a Heavy Casual overhaul that took place between 10th April and 20th May. The locomotive has been moved from the works to Crewe South depot to await a return to its home depot. Following another Heavy Casual overhaul the following year, it remained in traffic until 27th February 1967 working out of Wakefield depot, being scrapped by Arnott Young at Parkgate, Rotherham. *(1374)*

No 92031 at the head of a four-coach train near Charwelton on the Great Central's London extension. Near this location were water troughs, laid between the rails, so that steam locomotives could replenish their tenders whilst on the move. This was accomplished by the fireman winding a handle on the front of the tender which lowered a scoop into the water as the train passed over the troughs and the forward momentum forced the water up a pipe and into the tender. The use of water troughs enabled trains to travel non-stop over longer distances without the necessity of stopping to take on more water. Allocated to Annesley at the time, No 92031 would end its days at Newton Heath depot in January 1967. *(1308)*

No 92032 passes through Peterborough station with mineral wagons in tow – the first vehicle, possibly a former wooden bodied private-owner vehicle, certainly seems to have seen better days. The water tower is a reminder that at one time Peterborough was home to a number of locomotive depots: East – operated by the Eastern Counties Railway, 10th December 1846 until September 1885; New England – Great Northern, 1852 until total closure on 30th September 1968; Spital Bridge – Midland, 1872 until 1st February 1960; Station – Great Northern, 7th August 1850 until c1904; Water End – London & North Western, September 1885 until 8th February 1932. No 92032 entered traffic on 17th November 1954; initially allocated to March it moved to New England a few days later – this locomotive, along with Nos 92031 and 92033, may have been wrongly allocated due to the speed of transfer. With the run down of Eastern Region steam No 92032 moved to Birkenhead in July 1965 and was withdrawn in April 1967 to be recycled by Cashmores at Great Bridge that October. *(1384)*

The photographer has company on the embankment as No 92033 heads a down train of empty mineral wagons near Belgrave & Birstall, to the north of Leicester, on the Great Central main line on 30th August 1961. Today the southern terminus of the heritage Great Central Railway is in the vicinity, although not utilising the original station, with access off the A563 Leicester ring road. No 92033 arrived at Annesley depot in June 1957 staying until 1965 when it moved south to Banbury – it did not last much longer, being withdrawn from Northampton depot in September the same year. In 1947 a service of regular fast freight trains was inaugurated by the LNER between marshalling yards at Annesley, near Nottingham, and Woodford Halse. These trains, travelling over the GCR route, carried coal from the Nottinghamshire coalfield and steel products from Yorkshire and the North East destined for the London area or South Wales. This regular service gained a reputation for fast running and enthusiasts adopted the nickname of 'Windcutters' while enginemen generally referred to them as 'Runners'. From the 1950s until the end of through freight services, long trains of BR steel bodied mineral wagons would have been an everyday sight along the GCR. *(1386)*

Right: No 92035 stands shoulder to shoulder with LMS Class 4MT 2-6-0 No 43088, on shed at Peterborough's New England depot. Both carry 34E shed plates, dating the image from April 1958 when the shed was recoded from 35A, the 4MT arrived in December 1950, staying until March 1963. The 9F spent two periods at the shed, December 1954 until January 1959 and April 1959 to June 1963. The shed closed to steam in January 1965. No 92035 met its end at the hands of T. W. Ward's cutters in April 1966 at the company's Killamarsh yard. No 43088 emerged from Darlington Works in December 1950 being allocated to New England where it would have seen service over the Midland & Great Northern Joint line until the cross-country route closed in 1959. It was transferred to Staveley (Barrow Hill) in early March 1963 before arriving at Lostock Hall in August 1967, from where it was withdrawn that December. *(1409)*

Above: No 92036 is seen running south through Peterborough station at the head of a rake of mineral wagons. The railway history here is complex and readers are referred to *Rail Centres: Peterborough* by Peter Waszak (Ian Allan 1984) for the full story – the Eastern Counties, Great Northern, London & Birmingham, Midland and Peterborough, Wisbeach & Sutton Bridge companies all served the town. No 92036 was delivered to the local depot, New England, when new in December 1954 and remained there until September 1962. It returned to the depot for a final time in June 1964, and, following a short period in store, was withdrawn in early December 1964 after barely 10 years in traffic. The site at New England was large enough for a triangle to be constructed for turning locomotives, rather than the usual turntable. *(1416)*

Opposite Top: No 92037 arrived at Immingham in June 1963, from New England, and was put to work serving the coal industry. It is seen here in March 1966 with a rake of hoppers from Boldon Colliery that was originally operated by the Harton Coal Co – the coal industry had been nationalised in January 1947. It is recorded that over 120 miners lost their lives at Boldon during its operational life before its closure in 1982. No 92037 was put into store in January 1965, being withdrawn the following month and recycled by Drapers of Hull in June. With the amount of 'clag' being thrown around by the locomotives, washday must have been dreaded should the wind be in the wrong direction. *(1428)*

Opposite Bottom: No 92038 heads a class C freight train just to the north of Stevenage station on 9th September 1961. The first 10 wagons are 'Presflo' wagons introduced in the 1950s specifically for carriage of powdered cement but vehicles to this design were subsequently used to transport other powdered commodities. The original Stevenage station was built in 1850 by the Great Northern Railway, despite the apparent hostility towards the railway being built there at that time due to the inevitable decline it would cause to local coach businesses, which all ended shortly after the station was opened. In 1946, Stevenage became one of the first New Towns, which resulted in a new town centre; the station was replaced in 1973, the replacement being officially opened on 29th September 1973 by Shirley Williams MP. Allocated ex-works to New England depot, No 92038 remained there until June 1963 when it moved to Immingham. Withdrawal came from Langwith Junction in April 1965 with recycling taking place by Cashmores at Guide Bridge the following July. *(1438)*

Above: No 92039 runs past Bounds Green depot as it heads south with a rake of mineral wagons. The original depot dates from 1929 and today serves as a maintenance centre for London King's Cross services, operated by Hitachi it maintains AT300 units for London North Eastern Railway, Hull Trains and Lumo. The carriage hiding behind the locomotive, No E48018, was one of a batch of 28 suburban coaches built at Doncaster in July 1955 so was around five months old on 15th November when it was passed by No 92039. The carriage survived in passenger traffic until January 1975 whereas its contemporary was scrapped in February 1966 following withdrawal from Langwith Junction depot in October 1965. *(1442)*

Opposite Top: With a train of 'Presflo' cement wagons in tow, No 92040 is seen at Chaloners Whin to the south of York. Up until 1983 this was where the lines from Selby and Pontefract connected. In the 1970s, the National Coal Board (NCB) began development of a new underground mining complex in the area around Selby, North Yorkshire – the Selby Coalfield. Because of the risks to trains from mining subsidence, the NCB made the proposal in 1974 for the line to be diverted, and, following a planning inquiry in 1975, received consent in 1976, with the recommendation that the BR route be re-sited. This diversionary route for the ECML was constructed and paid for by the NCB. After its opening in 1983, ECML trains no longer called at or passed through Selby, instead leaving the former ECML at Templehirst junction and connecting with the former York & North Midland Railway line to York at Colton junction near Church Fenton. No 92040 entered service at New England in December 1954, being withdrawn from Langwith Junction depot in August 1965 and consigned to Drapers of Hull for disposal. *(1453)*

Opposite Bottom: No 92041 runs past the box at Retford South junction, heading east at the head of a class F freight consisting of unfitted stock. The tank wagons are marshalled away from the locomotive to reduce the fire risk. There were two locomotive depots at Retford, the Great Northern one being to the west of the GN station, closing on 14th June 1965, and the Great Central's Thrumpton shed to the east of the station on the south side of the Gainsborough line. Withdrawn from Langwith Junction in August 1965, No 92041 was recycled at T. W. Ward's Beighton yard that November. *(1463)*

Chapter 2: The class of 1955: Nos 92020-29, 92042-69

Crime of the Crostis?

In late 1950, the Government had directed the Railway Executive to reduce its coal consumption by 10,000 tons a year, and the decision was taken to equip an unspecified number of 2-10-0s with Franco-Crosti boilers. The Italian sponsors of this design claimed fuel savings of up to 20% when compared with conventional boilers. However, Riddles computed that a 9% saving per annum would save approximately 100 tons of coal per locomotive a year, enough to cover the conversion costs of the 10 locomotives; a royalty fee of £800 per boiler would be paid to the Italians. Readers are referred to books listed in the Bibliography where the reasons are discussed, but within five years the 10 conversions would be regarded as a fiasco. Having retained the smaller diameter boiler, when the locomotives were converted to 'normal' operation they were down-rated to 8F.

Below: No 92020 stands outside Crewe Works in ex-works condition illustrating the original configuration of the blast pipes and chimney. It was released to traffic on 18th May 1955, being allocated to Wellingborough. Two months later it was back in the works for a non-classified overhaul and stayed two months; not a good omen for a design that was designed to reduce operating costs. What is probably another member of the batch stands in front of No 92020, in grey undercoat, awaiting entry to the paint shop. Of the 10 converted Crostis, eight ended their days at Birkenhead Mollington Street – No 92020 being the first to be withdrawn in mid-October 1967, within three weeks the other seven were stood down. (1218)

Opposite Top: A clear view of the right-hand side arrangement showing the blast pipe arrangement, with the chimney being hidden behind the smoke deflector plate that was fitted to all locomotives, commencing in November 1955 following complaints from the footplate crews. The side chimney caused dirty conditions on the footplate with smoke entering the cab and the right-hand front window rapidly sooting up. Note the lamp columns being constructed from recycled bullhead rail. (1544)

Opposite Bottom: Creating its own smoke screen, No 92022 is seen in as-built condition at Ampthill on 20th September 1955 as it heads a train of empty mineral wagons along the Midland Main Line. Following the decision to rebuild the Crosti variant, it was stored for almost three years before being entering Crewe Works in April 1962; it returned to traffic in June, surviving in service until withdrawn from Birkenhead in November 1967 when the depot closed to steam. (1245)

Above: No 92025 is being serviced at Wellingborough on 4th May 1958. The shed was home to all 10 of the Crostis, and 65 other members of the class over the years. It closed to steam in October 1967, but the last 9Fs had departed in February 1964. Before its closure in 1984, Class 08 shunters, and Classes 25, 31, 45, 46 and 47 were stabled there. *(1298)*

Opposite Top: It's always nice to see the fruits of one's labours and the design staff at Brighton Locomotive Works were treated to a visit from No 92028 on 9th September 1955 where it is seen outside the works. Almost four years later the locomotive was placed into store in August 1959, entering Crewe Works for conversion to 'standard' condition that December. The works was one of the earliest railway-owned locomotive repair works, founded in 1840 by the London & Brighton Railway, and thus pre-dating the more well-known railway works at Crewe, Doncaster and Swindon. The works grew steadily between 1841 and 1900 but efficient operation was always hampered by the restricted site, and there were several plans to close it and move the facility elsewhere. Nevertheless, between 1852 and 1957 over 1,200 steam locomotives, as well as prototype diesel-electric and electric locomotives, were constructed there, before the eventual closure of the facility in 1962. *(1354)*

Opposite Bottom: After less than four years in traffic and a combined locomotive mileage of less than 800,000 miles, the decision was taken to revert to a conventional design with Derby Works allocated the task of producing the required drawings. No 92027 was placed into store at Wellingborough in April 1959, remaining so until 28th August 1960 when it entered Crewe for the preheater and side chimney to be removed. Its condition evokes thoughts that a few years later locomotives in this condition were on the way to the scrapyard. *(1474)*

No 92021 illustrates the group of 10 as rebuilt and is seen at Neath on 4th July 1964. Note that at this date there is no step beneath the smokebox to aid access. The image clearly shows the arrangement of the filler pipes for the sanding gear to the leading driving wheels; access from the running plate was not possible with the additional Crosti pipework. Only three 9Fs were ever allocated to Neath, Nos 92216, 92222 and 92225, from October 1963 – Nos 92216 and 92222 left for Southall in September 1964, 92225 had left for Newport Junction in May. *(1239)*

No 92026 was the last to remain in service in Crosti-condition and was never placed into store, it entered Crewe for conversion to standard layout in September 1959, being returned to Wellingborough. It is seen light engine at Chester Central on 1st September 1965 whilst allocated to Birkenhead Mollington Street. Withdrawn in November 1967, it was sold to G. H. Campbell in March 1968 and scrapped the following month in Airdrie. *(1315)*

Above: No 92042 was delivered to New England in January 1955 and 18 months later looks as if it was due for scrapping! It is seen at Potters Bar on 5th June 1956 at the head of a mineral train. It moved to Colwick in June 1963, then onto Langwith Junction in January 1965 before returning to Colwick in October the same year. Withdrawn a couple of months later, on 19th December 1965, it was consigned to J. Cashmore at Great Bridge before moving to T. W. Wards for scrapping at Beighton, Sheffield. The contraption to the left of the image between the first pair of tracks is an electric point motor. *E. Trotter (1467)*

Below: While waiting for the signal to be cleared at Cricklewood, No 92045 was photographed on 30th April 1955, already looking careworn – although the smokebox door has been cleaned – it was new to traffic on 9th February 1955, being allocated to Wellingborough. The following February it was reallocated to Toton, then onto Bidston before a final move to Birkenhead in February 1963. Withdrawn in September 1967, it become a casualty of T. W. Ward's Beighton scrapyard the following February. *A. G. Ellis*

British Railways Class 9F 2-10-0s

Below: Running tender first, No 92046 is at the head of a rake of ore wagons at Seacombe junction on 28th May 1958. The fleet of unfitted bogie wagons were built for J. Summers by Charles Roberts between 1952 and 1958 and carried iron ore from Bidston Docks, Birkenhead to British Steel's Shotton works. Nos 92045 and 92046 arrived at Bidston, initially on loan from Toton in May 1956; they and four other members of the class remained at the depot until February 1963 when all were transferred to Birkenhead. No 92046 survived in traffic until October 1967 and consigned to Newport for recycling in the South Wales furnaces. *(1496)*

Opposite Top: The inside of York shed with Nos 92049 and 92211 alongside an unidentified Class V2 2-6-2. No 92211 arrived at York depot in September 1963, departing for Wakefield in October 1966, possibly dating the image to this period. No 92049 was ex-works from Crewe on 10th March 1955, being allocated to Wellingborough, and arrived at its final depot of Birkenhead in February 1966. It was withdrawn in November 1966 and, following a period stored at Carlisle Kingmoor, was purchased by G. H. Campbell, Airdrie, where it was broken up in June 1968. *(1501)*

Opposite Bottom: With an age difference of over 60 years, former Lancashire & Yorkshire Railway Class 27 No 52218, dating from September 1893, stands alongside No 92050 that dates from August 1955. The image was probably taken at Crewe Works as No 52218 was one of the works shunters following its arrival in July 1953 – it survived until May 1962. No 92050 was ex-works at Crewe on 19th August 1955, and, initially allocated to Toton, it would be withdrawn from Warrington in September 1967 and scrapped in January 1968. An unidentified Stanier 2-6-4T tries to hide behind the 9F. *(1505)*

Opposite Top: Saltley-based No 92051 heads an express passenger service up the 1 in 38 Lickey incline. The locomotive was shedded at Saltley from June 1957 until November 1959 when it moved to Rowsley. The shed at Bromsgrove was the home of No 92079, illustrated on page 41, that supplied additional power for trains climbing the bank. No 92051 arrived at its final depot of Carlisle Kingmoor in December 1965, being withdrawn in October 1967 and sold to the Motherwell Machinery & Scrap Co at Wishaw where it was cut up in February 1968. *(1513)*

Opposite Bottom: Wellingborough-allocated No 92052 heads an express passenger train into London St Pancras on 26th March 1961, passing the iconic framework for the gasholders as it does so. The structures were built in the 1850s as part of Pancras Gasworks. The gasholders remained in use until the late 20th century and were finally decommissioned in 2000 to make way for redevelopment. When the regeneration of the area started, Gasholder No 8, together with 10, 11 and 12, was dismantled and shipped piece by piece to Shepley Engineers in Yorkshire. It took two years to restore Gasholder No 8, and in 2013 it returned to King's Cross and was rebuilt piece by piece in its new home on the banks of the canal in what is known as Gasholder Park. *(1516)*

Above: Doing what it says on the tin! No 92053 powers a rake of empty mineral wagons through the station at Bushey & Oxhey on 27th May 1961. The locomotive had arrived at Wellingborough shed in November 1959, where it remained until moving to Toton in September 1962. The London & Birmingham Railway first ran through here on 20th July 1837. A station was not initially provided, as the area was then sparsely populated, although in 1841 a station was provided in a similar red brick style to others along the route. London Underground trains served the station from 16th April 1917 until 24th September 1982. *(1525)*

Below: No 92054 running southbound near Kenton (Middx) at the head of a coal train sometime during 1956; its BR1C tender looks to be running on empty. New from Crewe Works on 21st September 1955, it was allocated to Toton, remaining there until March 1958 when reallocated to Wellingborough. It was withdrawn from Speke Junction in May 1968 and scrapped the following month. The coal may have been heading for the former LNWR power station at Stonebridge Park that provided electricity for the suburban services at that time or for some other industrial purpose – in the original image the lumps of coal appear too large for domestic use. *(1543)*

Opposite Top: Returning from permanent way duties, No 92055 is seen to the north of Standish Junction at the head of an empty ballast working in 1963. The vehicle behind the tender is a 'Shark', the code used by BR to describe all its ballast plough brakes. These were used to spread the ballast along the track prior to it being levelled. Continuing the name game the ballast hopper is a 'Herring' – all permanent way vehicles were named after fish. No 92055 was allocated to Toton at the time having arrived from Wellingborough in September 1962, before moving to Warrington in March 1965. It was withdrawn from Speke Junction in December 1967. *(1546)*

Opposite Bottom: No 92056 is seen at Lostock Hall on 12th October 1967 at the head of a rake of wagons containing soda ash, with little more than a month before it was withdrawn from service at Carlisle Kingmoor. At the time most soda ash was used to de-calcify or remove sulphurs and phosphates from some ferrous and non-ferrous ores. Today, soda ash is mined from trona deposits and is one of the world's most abundant natural resources. After it is processed, soda ash helps create a variety of products across many industries, such as building products, automotive, home, cleaning, energy, beverage, glass and bake ware, water treatment, oil and gas. *Arnold W. Battson*

Opposite Top: Proudly wearing its 18A, Toton, shed plate, No 92058 is seen light engine at Bedford in June 1962. Despite the fact that all trains should carry a head code, the locomotive carries no means of identification – there should be a lamp on the centre bracket above the buffer beam. It entered service at the depot on 13th October 1955, moving to Wellingborough in March 1958 before returning to Toton in September 1960. Withdrawal came at Carlisle Kingmoor in November 1967 with scrapping the following February by McWilliam's at its Shettleston yard. *Jack A. C. Kirke*

Opposite Bottom: At the head of an up freight consisting of a mixed rake of mineral wagons, no doubt carrying coal for the metropolis, No 92059 was photographed passing Hendon Aerodrome at 6.40pm on Saturday 5th May 1956. New to traffic at Toton in October 1954, the locomotive would remain there until moving to Wellingborough in March 1958; it ended up at Birkenhead in May 1965, being withdrawn in September 1966. Hendon Aerodrome was an important centre for aviation from 1908 to 1968, and was briefly active during the Battle of Britain, but for most of World War 2, it was mainly used for transport activities, and flying dignitaries to and from London. Late in 1968, when two of the three runways had been removed, a Blackburn Beverley was flown in to be an exhibit at the new RAF Museum – this was the very last aircraft to land at Hendon. The RAF station finally closed in 1987 with the museum situated on the south east side of the site. *Arthur Carpenter*

Above: You can almost feel the power as No 92060 waits for the signal, near the overbridge, to clear at Tyne Dock during March 1965. The Westinghouse pumps, mounted on the running board, are shrouded in steam and were to provide air for the operation of the doors on the new 56-ton capacity hoppers specially designed for the iron ore traffic. The locomotive was the first of seven allocated from the 1954 building programme for use on the service and all were allocated to Tyne Dock. Nos 92060-66 and 92097-99 (from the next build programme) were fitted with air pumps. *Jack A. C. Kirke*

Opposite Top: The 4.15pm Tyne Dock Bottom-Consett iron ore train is nearing Tyne Dock Bank Top on 23rd July 1962 behind No 92063. New from Crewe towards the end of 1955, the first group were allocated to Wellingborough as temporary replacements for the Crostis that were being modified at the time. The air pumps were installed on the fireman's side between March and May 1956 before the locomotives were reallocated to the North Eastern Region, see image of No 92098, page 46. With one exception, No 92065, the Tyne Dock locomotives were withdrawn between May 1965 and November 1966. No 92063 was taken out of service in November 1966, almost 11 years to the day after it emerged from Crewe Works, and scrapped by T. J. Thompson at Stockton-on-Tees the following April. *Brian Wadey*

Opposite Bottom: Just to the north of Chester-le-Street was Ouston junction where the North Eastern's main line to Newcastle connected with the line running westwards to Consett, the destination of this train of iron ore from Tyne Dock on 13th April 1957. No 92064 is to the south of the junction and the Birtley Grange Brick & Tile Works can be seen to the left of the locomotive. No 92064 arrived at Tyne Dock depot from Crewe in December 1955; it was loaned to Wellingborough and Toton depots over the next few months, returning permanently to Tyne Dock in May 1965 where it would spend the next 10 years. Withdrawn in November 1966, it would be scrapped the following April in Stockton-on-Tees.

Above: The Great Central main line was engineered for fast running trains, and this included heavy freight trains, although in this case No 92069's train seems only to consist of a couple of bogie wagons carrying steel bars on 30th August 1961, the rest of the train being hidden behind the 'box. It is passing through Belgrave & Birstall station that was opened on 15th March 1899 serving the villages of those names. Taken over by the London & North Eastern Railway at the Grouping, the route became part of the Eastern Region when nationalised. The route was transferred to the London Midland Region in the early 1960s, which closed the GCR's 'London Extension' on 4th March 1963. In 1991 Leicester North station was opened immediately to the south of Belgrave & Birstall station by the heritage Great Central Railway and is the southern terminus of the line. It is representative of the 1960s when the line was under the control of BR London Midland Region.

Below: Trying to be a 'stealth' locomotive, No 92077 runs through Preston station in December 1967. It left Crewe Works on 13th March 1956, being allocated to Toton. Arriving at Carnforth shed in March 1967, it survived until withdrawal in June 1968. It was sold to G. H, Campbell of Airdrie but was recycled by Clayton & Davie of Dunston, Gateshead, in October 1968. The 20 ton brake van, No B951816, was part of Lot 2350 built at Darlington in 1952. Preston station is roughly halfway between London Euston and Glasgow Central (206 miles from London and 194 from Glasgow). The North Union Railway opened a station on the site in 1838. It was extended in 1850, with new platforms under the separate management of the East Lancashire Railway, and by 1863 London-Scotland trains stopped here to allow passengers to eat in the station dining room. The current station was built in 1880 and extended in 1903 and 1913, when it had fifteen platforms. A free buffet for servicemen was provided during both World Wars. The East Lancashire platforms were demolished in the 1970s as connecting lines closed. *(1331)*

Opposite Top: Using the through line at Wellingborough Midland Road, No 92078 hauls a down freight consisting of empty mineral wagons through the station at 12.50pm on the 28th July 1956; the service was a class F consisting of unbraked wagons. The locomotive was new to Toton on 21st March 1956 and withdrawn in early May 1967; it was scrapped the following January by T. J. Thompson at Stockton-on-Tees. The water tank looks like one of those ready-made examples that occupied many model railways. *A. Lathey*

Opposite Bottom: Following withdrawal in May 1956 of the Midland Railway's heavyweight four-cylinder 0-10-0, BR No 58100 – otherwise known as 'Big Bertha' – which was only the second ten-coupled locomotive to enter service in the UK – was replaced by No 92079 banking trains up the Lickey incline. The incline, from Bromsgrove to Blackwell summit, was two miles of 1 in 38 and the majority of trains would require banking. The 9F inherited the electric headlamp fitted to the MR locomotive, spending five days at Derby for the work to be carried out. Its original flush-sided BR1C tender was replaced with an inset BR1G variant, as half its mileage would be run in reverse. At some date a rectangular aperture was cut in the left hand side of the tender to aid coaling on Bromsgrove shed, this can be seen immediately behind the cab in the original image. No 92079 remained at Bromsgrove until moved to Birkenhead in October 1960. Its duties on the Lickey were shared with GWR-designed '94XX' 0-6-0PTs. The headlamp was not always used at night as some drivers claimed the train locomotive crew did not pull their weight if they knew what the banking locomotive was. *J. Davenport*

Opposite Top: Awaiting its coat of black, No 92080 stands in the yard at Crewe Works on 5th April 1956. Nine days later it entered traffic, being allocated to Toton. It remained there for two years before moving on to Wellingborough, Kettering, and Newton Heath before arriving at its final depot, Carlisle Kingmoor, in August 1966. It was withdrawn in early May 1967 and, following storage, arrived at the Shettleston yard of J. McWilliams for disposal. *Ron F. Smith*

Opposite Bottom: No 92083 is seen under the coaling tower at Cricklewood shed on 7th August 1957, having worked in from Wellingborough. It moved around a fair bit after arriving at Annesley in November 1960 before arriving at Birkenhead in May 1965 from where it was withdrawn in mid-February 1967. It was another of Albert Draper's victims, this time on 31st July 1967. The Midland Railway opened Cricklewood depot in 1892, closing to steam on 14th December 1964, but it continued to be used as a diesel stabling point for a few years. The site has been used for Brent Cross West station – the first major new main line station in London in over a decade – with the depot being moved to a larger site slightly to the south. *Ron F. Smith*

Above: With at least three ex-LNER parcels vans hanging on the draw bar, No 92088 runs southbound into Banbury station beneath Bridge Street overbridge with a class C service of express-passenger rated stock. The original GWR Banbury Bridge Street station opened on 2nd September 1850, some four months after the Buckinghamshire Railway (L&NWR) opened its Banbury Merton Street terminus on the line from Bletchley via Verney Junction. To the north of the station the Great Central Railway arrived in 1900 connecting the town to the GC main line. At its height there were six routes concentrated on the town, with easy access to a seventh. Delivered new to Doncaster shed in October 1956, No 92088 survived until April 1968, latterly operating from Carnforth depot. It met its end at the hands of Albert Young's cutters at Dinsdale the following October.

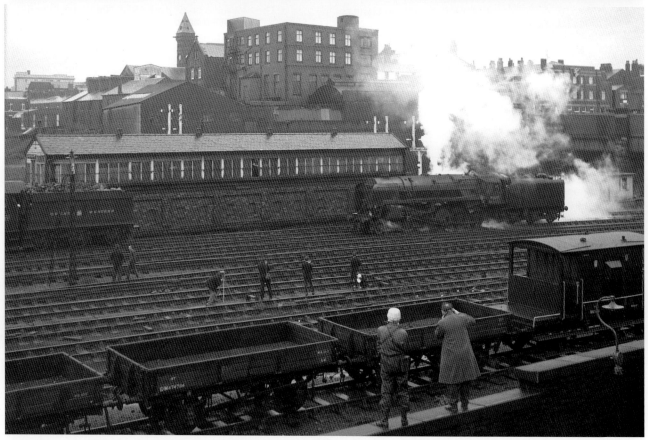

Opposite Top: With the properties of Western Boulevard backing onto the railway, No 92090 is seen to the south of Upperton Road bridge as it heads a class H freight service past Leicester South Goods in July 1957. At this date the locomotive was allocated to Annesley depot having arrived from Doncaster the previous March; it was withdrawn from Birkenhead in the middle of May 1967. Leicester's Great Central Railway station was opened on 15th March 1899, being part of the company's London extension linking Nottingham with Marylebone in London. The railway crossed through built-up Leicester on a Staffordshire blue brick viaduct, incorporating a series of girder bridges. In a detail typical of the high standards to which the London extension was built, the abutments of the girder bridges that crossed public roads were lined in white-glazed tiles to increase the level of light under the bridges. In total the viaduct was in excess of a mile and a half in length and it was upon this that Leicester Central station would be constructed. At the time of construction, the station was the largest single building to be erected in Leicester. On 3rd September 1966, the line ceased to be a trunk route with the withdrawal of services to Sheffield and Marylebone, leaving Leicester Central operating a sparse DMU local service to Nottingham and Rugby. *Canon Alec George (AG613)*

Opposite Bottom: It is the 14th October 1967, and the Locomotive Club of Great Britain ran its 'Castle to Carlisle' rail tour using GWR 'Castle' No 7029 *Clun Castle*, by now privately owned, and No 92091. The 9F took the train from Liverpool Exchange to Preston where the 'Castle' took over for the run to Rome Street junction (Carlisle) where an Ivatt 2-6-2T took over for a trundle around the area. The 'Castle' took the train back south to Hellifield where the 9F took over for the return to Liverpool. The scene here is at Preston with the locomotive change over taking place – no worries about hi-vis clothing or health and safety concerns in those days! *Arnold W. Battson*

Left: No 92093 is seen beneath the ash tower at Carnforth, in the company of a 'Britannia' Pacific, wearing its Carlisle Kingmoor shed plate. The depot was completed by the LMS in 1944 on the site of the former Furness Railway facility. BR thus inherited an almost brand new depot, which was larger than was really required, allowing the closure of a number of other local and older or less efficient sheds, and to keep the shed open longer than many when the decision to modernise traction to electric and diesel came. Closure came on 5th August 1968 with the end of steam on the main line. Targeted as part of a preservation scheme, when this failed it was developed as a visitor attraction – Steamtown Carnforth. Today, closed as a museum, it acts as the major national operational base of West Coast Railways using mainly former Southern and LMS steam locomotives along with a collection of more modern – well 1960s and 1970s vintage – motive power. The ash tower and coaling tower, behind the photographer, still stand but are not in use.

Opposite: The fitment of air pumps was mentioned earlier with No 92063, see page 38, and conveniently these two images show the arrangement on both sides of the locomotives. No 92098 is seen on the turntable at Tyne Dock on 3rd August 1958 showing both air pumps and additional pipework on the fireman's side. The driver's side of No 92099, on 17th July 1965, shows the addition of an air receiver below the running board between the third and fourth driving wheel sets. By the time the depot was closed on 9th September 1967 the buildings were in derelict condition; the site was cleared in the early 1970s. No 92098 was withdrawn from Tyne Dock shed in late July 1966 with 92099 following around the end of August/early September. Both were dispatched to Drapers of Hull for disposal that occurred in October and November respectively.
Arnold W. Battson

Above: The fireman has spotted the cameraman as No 92100 takes its train north through Derby station that lies 127 miles 68 chains (205.8km) north of London St Pancras. The station is situated to the south east of Derby city centre, and is close to the west bank of the River Derwent, being opened by the North Midland Railway on 11th May 1840. New to traffic at Toton in July 1956, No 92100 was shunted around various LMR depots, spending two periods at Leicester Midland, before ending its days, as with so many classmates, at Birkenhead. Withdrawn in May 1967, it was sold to G. Cohen at Kettering for scrapping. Although Derby was home to the Midland Railway's locomotive works, very few of the class visited for repairs.

Opposite Top: Following a Heavy Casual overhaul, No 92104 is seen at Crewe Works on 1st October 1961. It is recorded that the locomotive was at the works from 25th August until 31st October, so if the date of the photograph is correct it stood around for a month before heading back home to Leicester Midland depot. On the original image it can be seen that the back plate of the tender step has the locomotive number stencilled on, ensuring it was connected to the correct locomotive before returning to traffic. No 92104 was transferred to Westhouses in April 1962, then moving via Speke Junction to Birkenhead from where it was withdrawn in February 1967. Sold to Albert Draper of Hull, it was cut up on 31st July 1967, another of the 37 to be broken up in the yard. *Arnold W. Battson*

Opposite Bottom: Wearing its 15B Kettering shed plate, No 92106 is seen shunting at Brimscombe on 26th September 1964 shortly before it was reallocated to Leicester Midland in October. The station here was opened on 1st June 1845 on what is now the Golden Valley Line between Kemble and Stroud in Gloucestershire. This line was opened in 1845 as the Cheltenham & Great Western Union Railway from Swindon to Gloucester. The station opened three weeks after the general opening of the line, originally as 'Brimscomb'. The station was renamed as 'Brimscomb near Chalford' in June 1865 and finally to Brimscombe on 2nd August 1897. Closure of the station came on 2nd November 1964 following the withdrawal of local stopping passenger services on the line. Goods traffic had ceased earlier on 12th August 1963. No 92106 has left its train on the running line and has shunted a short rake of mineral wagons into the goods yard. It would survive, by now allocated to Birkenhead, until withdrawn in late July 1967 and scrapped that December by T. J. Thompson at Stockton-on-Tees. *James L. Stevenson*

Below: In 1847, the GWR opened the Gloucester & Cheltenham Loop line that completed the triangle junction to the east of the station. This allowed GWR trains to avoid the reversal at Gloucester, but so as to allow GWR passengers to access Gloucester, a link line was built to a station on the loop called the Gloucester T station. Carriages were detached from trains at the T station, turned on turntables and taken into the main Gloucester station. This operation was not very successful and so was abandoned, along with the loop line, in 1851. In 1901, the Cheltenham Loop – now known as the Gloucester avoiding line – was re-instated, primarily for goods traffic, but also for passengers from 1908. It is on this section of line that No 92108 is seen hauling a westbound freight along the loop between Engine Shed and South junctions, Gloucester on 17th October 1965. New to traffic in October 1956, and initially sent to Wellingborough, by now the locomotive was allocated to Birkenhead from where it was to be withdrawn in November 1967. *Peter Kerslake*

Opposite Top: The leading wagon, No B852849, is branded as a 'Shocvan', these were special wagons used to transport fragile goods that were at risk of being broken as a result of heavy shunting or other jolts whilst in transit; this was especially important for eggs and pottery. The van bodies were about 10in (254mm) shorter in length than the chassis and were mounted to the chassis via springs that absorbed some of the shock of sudden jolts. No B852849 was one of a large batch of vehicles constructed at Darlington in 1955, in February 1967 the wagon carries advertising labels for Earles Cement and 'Smoke Packs From Metal Box' – neither of which are really fragile! No 92110 stands at the south end of Preston station as it heads an up freight service. Allocated to Carlisle Kingmoor at the date of the photograph, the locomotive would survive in traffic until December 1967 when it would be condemned, being sold for scrap the following March. *Arnold W. Battson*

Opposite Bottom: No 92112 running southbound through Elstree & Borehamwood station at the head of a rake of mineral wagons during 1958 – the single lamp indicating a class J service (through mineral or empty wagon train). On 22nd June 1863, the Midland Railway (Extension to London) Bill was passed: 'An Act for the Construction by the Midland Railway Company of a new Line of Railway between London and Bedford, with Branches therefrom; and for other Purpose'. Situated north of the Elstree Tunnels, the station was built by the Midland Railway as simply 'Elstree', opening on 13th July 1868, when it built its extension to London St Pancras. It was renamed Elstree & Boreham Wood the following June and then Elstree & Borehamwood on 21st September 1953. Allocated new to Cricklewood in November 1956, No 92112 remained at the depot until reallocated to Wellingborough in April 1959. It was sent to Birkenhead in April 1966 from where it was withdrawn in November 1967. It was sold to G. H. Campbell of Airdrie where it was cut up in March 1968.

Above: Barton, to the north of Preston, is the location of this image of No 92114 hauling a train carrying rails originating from the British Steel Corporation's site at Workington. Workington rails were widely exported and a common local phrase was that Workington rails 'held the world together'. Originally made from Bessemer steel, but after the closure of Moss Bay Steelworks, steel for the plant was brought by rail from Teesside. The plant was closed in August 2006. No 92114 entered traffic on 17th November 1956, being allocated to Westhouses depot, and, as with many of the class, it ended its days at Carlisle Kingmoor, having arrived in May 1965 from where it was withdrawn in early November 1967. *Arnold W. Battson*

Above: Running south through Tapton junction towards Chesterfield Midland station, No 92116 is at the head of a class F freight consisting of unfitted mineral wagons. It carries an 18B, Westhouses, shed plate where it had arrived new from Crewe Works in December 1956; it moved to Wellingborough in March 1963. Tapton junction was on the former Midland Railway route where the line to Sheffield and Staveley diverged. Chesterfield was also served by the Lancashire, Derbyshire & East Coast (later Great Central) Railway's station that closed to passengers on 3rd March 1963 and to freight services on 11th September 1967. Withdrawn from Warrington depot in November 1966, No 92116 was sold to Drapers of Hull for recycling which took place on 12th June 1967.

Below: No 92118 is seen at Banbury in September 1965, this had been home to the locomotive from July to November 1964 so it had worked in from Tyseley depot at the time of the photograph. Its final shed was Carnforth where it had arrived in November 1966 to be withdrawn in May 1968, as one of the last to remain in service. No 6952 *Kimberley Hall* was allocated at Banbury for a second time between November 1962 and October 1965 when it moved to Tyseley. The GWR shed was opened on 6th October 1908, it was coded 84C as a Western Region depot, becoming 2D when taken over by the London Midland Region in September 1963, closing on 3rd October 1966. *Jack A. C. Kirke*

Chapter 4: The class of 1957: Nos 92093-96, 92119-62, 92168-70, 92178-83

No 92096 was delivered from Swindon Works, as the last member of Lot 421, to Annesley depot on 2nd April 1957. It is seen here with a class H freight service, at Beattock summit, on 20th July 1966 whilst allocated to Carlisle Kingmoor, having been officially stored since October 1965 – although it was reported as seeing service at Derby in December – possibly reinstated for the additional Christmas traffic. Arriving at Kingmoor in January 1966, it stayed in traffic for a little over a year, being withdrawn in February 1967. The photographer was not alone, note the pair further along the siding complete with tripod!
James L. Stevenson (1308)

Opposite Top: Chester Central on 22nd August 1966 with No 92119 running tender first through the station.
On 23rd September 1840, the first station at Chester was opened by the Chester & Birkenhead Railway (CBR). One week later, on 1st October 1840, the Grand Junction Railway (GJR) opened a separate station. Neither station was open for long, due to the inconvenience of transferring goods and passengers between them. They were replaced by the new joint station at the junction between the CBR, GJR and Robert Stephenson's new Chester & Holyhead Railway that started at the joint station. The station was designed by the architect Francis Thompson, and constructed by Thomas Brassey. The engineer C. H. Wild designed the train shed. Construction of the station commenced on 1st August 1847, with the foundation stone being laid by Brassey. It was officially opened, exactly a year after construction began. No 92119 was delivered from Crewe Works on 12th January 1957, being allocated to Westhouses where it remained until moving to Cricklewood two years later. It was transferred to Carlisle Kingmoor in August 1967 and withdrawn virtually immediately. *James L. Stevenson*

Opposite Bottom: No 92122 rolls into Gloucester Central over London Road on 16th February 1965 with a train containing steel coil from Newport to Birmingham. The climb to gain height over the road is obvious. At this date No 92122 was allocated to Birkenhead; it had been delivered new to Wellingborough in February 1957 from Crewe Works, moving to Leicester Midland depot in March 1960. Its final move, to Birkenhead, took place in April 1965 with withdrawal in November 1967, to be scrapped in Airdrie by G. H. Campbell. The foot crossing is for the use of station staff, note the sign to the right of the image – 'Passengers are requested to cross the lines by means of the bridge'. *Peter Kerslake*

Above: No 92123 pokes its nose out of the Midland Railway-built roundhouse at Wellingborough that opened in 1868. The roundhouse was refurbished by the LMS in the early 1930s and demolished in July 1964. A second roundhouse, to the south of the original, was opened in 1872. Wellingborough shed closed to steam on 13th June 1966 with final closure as a diesel depot in 1984. The southern of the two roundhouses remains standing with a recently built road through the site called Roundhouse Road. No 92123 arrived at Wellingborough in February 1957, remaining until relocated to Leicester Midland in March 1960. A final move to Birkenhead occurred in April 1965 with condemnation in October 1967. *Ron F. Smith*

Opposite Top: Knighton South junction with No 92124 passing with a class C freight service on 28th August 1961. The triangular junction was situated to the south of Leicester London Road station on the former Midland Counties Railway, with the Midland Railway line to Swannington heading off to the west. Following closure of the former Great Central line that also served Leicester, BR constructed a connecting line to Braunstone Gate goods yard in 1965. In 1973 Vic Berry established his scrapyard on this site just south of the former Leicester Central station. He focused initially on breaking up redundant carriages and goods stock. The first locomotives did not arrive until 10 years later in April 1983 when three Class 76 electric locomotives arrived for breaking up. The yard is best known for scrapping large quantities of Class 25 and Class 27 diesel-electric locomotives. This led to the infamous 'stack' of Class 25 and 27 locomotives that reached their peak in 1987 with 30 examples stacked. Following a serious fire in March 1991 the site was closed. No 92124 was allocated to Wellingborough at the time of the photograph, its final shed was Warrington from where it was withdrawn in December 1966.

Opposite Bottom: Seen at its home shed of Croes Newydd on 21st August 1966, No 92125 stands in front of No 92135; both locomotives had arrived at the depot that May from Saltley. The Great Western Railway opened the depot at Croes Newydd, Wrexham, in 1902 – its facilities included a coal stage, with water tank over, and a repair shop. It was reroofed in 1924 and closed by BR on 5th June 1967. No 92125 left Crewe Works on 9th March 1957, being allocated to Wellingborough and moving via Kettering to Saltley. It was withdrawn in December 1967 from Carlisle Kingmoor. Meanwhile No 92135 had left Crewe Works on 11th June 1957 and withdrawal took place from Wakefield on 8th June 1967. *James L. Stevenson*

Above: Wellingborough allocated No 92126 is seen with a mixed freight in tow near Harpenden. Once in open countryside, today housing backs on to the line. The station is the second built in the town, by the Midland Railway in 1868 on its extension to St Pancras – nothing remains of the original station buildings. Although located on Station Road, the road is actually named after the first station, Harpenden East, which was built in 1860 and closed in 1965. A branch line, built by the Hemel Hempstead Railway in 1877, known as the Nickey Line but operated by the Midland, formerly diverged from the main line north of the station. The intention had been to meet the LNWR at Boxmoor, however the section from Hemel Hempstead never had a passenger service. In 1886, a south curve was added to the junction allowing passengers to join the London trains at Harpenden rather than Luton. The branch was closed in 1979, but the route remains in use as a cycleway, passing under the M1 in a tunnel. New to traffic on 18th March 1957, No 92126 entered traffic at Wellingborough; moving to Kettering in February 1964, it was withdrawn from Warrington in August 1967 and disposed of by J. McWilliam at Shettleston. *Nigel Lester*

Above: No 92129 powers a mixed consist of class C, express passenger-rated stock, southbound along the four-track main line at Wellingborough in June 1961 whilst shedded at Cricklewood – over the short lives of the class, 11 locomotives were allocated to the London depot, No 92129 was one of the last three to leave in October 1961 when it moved to Annesley. It remained in service until June 1967 when it was despatched to Wishaw for scrapping the following November. Wellingborough depot buildings can be seen behind the train. *Jack A. C. Kirke*

Below: No 92130 leads an excursion train out of Rotherham Masborough station, the station was opened by the North Midland Railway as Masbrough on 11th May 1840, being renamed Rotherham Masborough in 1908. It closed on 3rd October 1988 when traffic was transferred to Rotherham Central. The line leading off to the left led to the Midland Iron Works, incorporated in 1865 and by 1937 the company was part of the Thomas W. Ward group – a name that will be familiar to readers of this book. A boiler explosion at the works on 3rd December 1862 resulted in the deaths of seven men with 25 injured. Primarily ship-breakers, the company recycled a large number of warships over the years, civilian vessels were also handled with the 1938-built RMS *Mauretania* meeting its fate at Inverkeithing in 1965. Following its withdrawal in June 1966, No 92130 met its demise at the Wishaw yard of the Motherwell Machinery & Scrap Co. *(4096)*

Above: New from Crewe in April 1957, No 92131 was allocated to Saltley where it was to remain for a few weeks before moving to Toton. It was virtually brand new when photographed as it's carrying a 21A shed plate. Early 1960 saw a move to Westhouses before moving to Speke Junction with a final move to Birkenhead from where it was withdrawn in September 1967. The Midland Railway opened its depot at Saltley in 1868 that was extended in 1876, and again in 1900. The first two adjoined roundhouses were re-roofed in concrete by the London Midland Region in 1948 with the final 1900-built roundhouse in glass and steel being completed in 1951. The depot closed to steam on 6th March 1967, being virtually demolished and replaced by a three-road diesel depot. *E. A. Elias*

Below: No 92132 at the head of a rake of 11 bogie coal hoppers, and brake van, is seen at South Kenton on 29th July 1961. The Birmingham Carriage & Wagon Works constructed the hoppers for the LMS in 1929. The batch of 30, Nos 189301-330, were used on the daily Toton-Stonebridge Park power station service and survived until 1967. No 92132 entered traffic on 29th April 1957, being allocated to Saltley before moving to Wellingborough that December. It spent its last three years moving round various LMR depots before being withdrawn from Carlisle Kingmoor in October 1967. *(4100)*

Opposite Top: Saltley-allocated No 92134 has been in the wars having suffered a heavy shunt resulting in a bent buffer beam, behind it stands an unidentified Hughes/Fowler 'Crab' 2-6-0. No 92134 underwent a Heavy Intermediate overhaul at the Southern Region's Eastleigh Works between 22nd July and 8th September 1964. Following withdrawal it was consigned to Woodham Brothers, Barry, for scrapping, but has survived to be the only single-chimney 9F in preservation with the final stages of its restoration being covered in the Channel 5 TV documentary 'The Yorkshire Steam Railway'. *(4105)*

Opposite Bottom: No 92136 rolls an express passenger service through Beauchief station, to the south of Sheffield, whilst it was allocated to Saltley depot. It was one of only a handful of 9Fs to be allocated to a single depot; it arrived new from Crewe in June 1957 and survived until October 1966. The Midland Railway opened the line in 1870, with the station here opening as Abbey Houses on 17th February, being renamed Beauchief on the 1st April the same year. A further change took place on 1st May 1874 when it became Beauchief & Abbey Dale – a final amendment to Beauchief taking place on 19th March 1914. At opening it had two platforms, but this was increased to four with the widening taking place between 1901 and 1903. Designed by the Midland Railway's company architect, John Holloway Sanders, it was closed to passengers on 2nd January 1961, and to goods traffic on 15th June 1964. *(4113)*

Above: No 92137 emerged from Crewe Works in June 1957, being allocated to Saltley before moving to Croes Newydd, Wrexham, in August 1966. It is seen here as it rolls a rake of mineral wagons through Gloucester Central station. The railway development at Gloucester was very complex involving four different railway companies and five distinct railway stations; after several amalgamations the Great Western and Midland railways were the operators. Between 1914 and 1920, the GWR station was expanded with a second long platform north of the running lines, two centre tracks for through movements and bay platforms. The two main platforms were also split in two with a scissors crossing in the middle. In 1951, the former GWR station was renamed Gloucester Central with the Midland station becoming Gloucester Eastgate to avoid confusion. A sign that the end of steam traction is near is the appearance of a Brush Type 4 (later Class 47) in the distance. As a point of interest, at 1,977ft 4in (602.69m), Gloucester has the second-longest platform in the UK. *(4114)*

Opposite Top: The quality of water has caused problems to No 92138 as the amount of scale on the boiler cladding demonstrates at Nottingham shed on 27th September 1963 – unless of course a driver had parked it beneath a limestone hopper! No 92138 had entered traffic at Saltley, having entered traffic from Crewe Works on 25th June 1957. Its one and only reallocation took place in August 1966 when it was moved to Speke Junction depot, from where it was withdrawn in early July 1967 to be scrapped by Birds Commercial Motors of Long Marston in February 1968. The first Midland Railway roundhouse in Nottingham was opened in 1868, being adjoined by two additional roundhouses in 1877 and 1893. The depot was closed on 4th April 1965 and subsequently demolished. *Richard Snook*

Opposite Bottom: In charge of a class H, through ballast or freight, service, No 92140 is seen crossing the junctions to the north of York station. The lines heading off to the left are the routes to Knaresborough and Darlington, those to the right lead to Scarborough and Market Weighton (the latter closing in 1965). York station is a key junction approximately halfway between London and Edinburgh, the first station being opened by the York & North Midland Railway in 1839; with a complicated history, readers are referred to *Rail Centres: York*, Ken Hoole, Ian Allan Ltd, 1983, ISBN 9780711013209, for a full history. No 92140 was allocated new to Peterborough's New England depot in July 1957, moving to Langwith Junction in January 1964 from where it was withdrawn in April 1965. It was delivered to J. Cashmore, Great Bridge, for scrapping in July 1965.

Below: No 92141 takes the centre road through Hornsey station whilst allocated to Doncaster; it had arrived there from New England depot in September 1962 only to return in March 1963. It was withdrawn from Colwick on 18th December 1965 and recycled by T. W. Ward the following April. The locomotive carries a class C head code indicating that the train is made up from express passenger-rated stock. The station at Hornsey was opened on 7th August 1850 by the Great Northern Railway, the same day that the main line between Peterborough and London (Maiden Lane) was opened. Under plans approved in 1897, the station was to be served by the Great Northern & Strand Railway, a tube railway supported by the GNR that would have run underground beneath the GNR's tracks from Alexandra Palace to Finsbury Park and then into central London. *(4128)*

Opposite Top: Seen passing northbound through Huntingdon on 24th October 1964 at the head of a class C, through ballast or freight, service is No 92142. Delivered new to Peterborough New England depot from Crewe Works in July 1957, it was the locomotive's only home. Following withdrawal in February 1965 it was initially stored at New England, then Colwick, before going to the Kettering yard of G. Cohen for scrapping in April 1966. Despite the fact that Huntingdon has expanded over the years, this section of line is still in the countryside, although dominated by the electrified four-track main line.
David Birt

Opposite Bottom: A 9F sandwich at Doncaster as No 92145 heads a class F, express freight with unfitted stock, northbound, having passed beneath Balby Road Bridge. To the left an English Electric Type 3 (later Class 37) heads a rake of express passenger-rated stock, whilst to the right a sister locomotive waits in the yard with a rake of coaches for a following northbound service. No 92145 entered traffic at Peterborough New England in August 1957; moving to Langwith Junction in January 1965, it was withdrawn from Immingham in February 1966. Doncaster was home to the Great Northern Railway's locomotive works; known as 'the Plant' it was established in 1853, replacing previous works in Boston and Peterborough. Until 1867 it undertook only repairs and maintenance, after the Grouping in 1923 it built locomotives like *Flying Scotsman* and *Mallard*. In 1957, BR Standard Class 4 No 76114, the last of over 2,000 steam locomotives, was completed. Carriage building ceased in 1962, but the works was modernised with the addition of a diesel locomotive repair shop. Under British Rail Engineering Ltd, new diesel shunters and 25kV electric locomotives were built, plus Class 56 and Class 58 diesel-electric locomotives – the works closed in 2007. *Ian Strachan*

Above: Another of the class delivered new to Peterborough's New England depot, this time in September 1957, No 92147 is seen running light engine northwards through York station, no doubt having worked in from its home depot. In March 1963 it was transferred to Immingham depot from where it was withdrawn in April 1965, being scrapped by J. Cashmore at Great Bridge in August 1965. Of interest to the modeller is the identification arrangement used for the points and signals as an aid to the Signalling & Telegraph staff responsible for maintenance.

Opposite Top: From just north of Helpston the Midland Railway's Syston & Peterborough line ran to the west of the Great Northern Railway line south to Peterborough. At Werrington junction the GNR's line from Boston and Spalding comes in from a North-Easterly direction. No 92148 heads north along the former GNR line in June 1963; it was allocated to the nearby New England depot from delivery in September 1957 to June 1960 and September 1961 to March 1963 when it was sent to Doncaster. It was withdrawn from Colwick depot on 19th December 1965 and scrapped by T. W. Ward at Beighton, Sheffield, the following April. *Jack A. C. Kirke*

Opposite Bottom: No 92149 is seen at Potters Bar in June 1963 at the head of a train of APCM cement wagons – originally nicknamed 'Silver Queens' due to the use of aluminium for the bodies. Built in two batches by Gloucester C&W (1961) and Metropolitan-Cammell Ltd (1963-65), they were one of the few successful designs to be built of aluminium, thereby giving a low tare weight and high gross load/tare ratio. The trains ran from Cliffe cement works in Kent to Uddingston, Glasgow, using New England 9Fs on the East Coast main line. No 92149 left Peterborough for Langwith Junction in January 1965 from where it was withdrawn at the end of June, being consigned to Ward's Beighton scrapyard. *Jack A. C. Kirke*

Above: New to traffic on 5th October 1957, No 92150 was allocated to Westhouses where it would remain until February 1959 when it was transferred to Saltley. It is seen here at Banbury during 1965 during its second stint at Saltley from June 1964 to October 1966. Withdrawn from Wakefield in April 1967, it was stored at Warrington before being consigned to Albert Draper in Hull who scrapped it on 27th November 1967. The leading two wagons are vacuum-braked 'Covhops' – built at several BR workshops including Derby in the 1950s and Ashford in the early 1960s to transport sand and soda ash amongst other commodities. A number used for sand traffic were branded 'BIS Sand for Rockware Glass Ltd'. *Jack A. C. Kirke*

Opposite Top: Saltley-allocated No 92151, minus its smokebox number plate, heads a southbound freight through the former Great Western Railway's station at Birmingham Snow Hill on 5th March 1966. Snow Hill was once the main station of the GWR in Birmingham, and at its height it rivalled New Street station, with competitive services to destinations including London Paddington, Wolverhampton Low Level, Birkenhead Woodside, Wales and south west England. The station has been rebuilt several times since the first station at Snow Hill, a temporary wooden structure, was opened on 1st October 1852; it was rebuilt as a permanent station in 1871, and then rebuilt again on a much grander scale during 1906-1912. The electrification of the main line from London to New Street in the 1960s saw New Street favoured over Snow Hill, which saw most of its services withdrawn in the late 1960s. This led to the station's eventual closure on 6th March 1972, and demolition five years later. After fifteen years of closure a new Snow Hill station, the present incarnation, was built; it reopened on 5th October 1987. No 92151 was reallocated to Birkenhead in November 1966, being withdrawn in mid-April 1967 and consigned to Drapers in Hull for scrapping. *Brian Wadey*

Opposite Bottom: With a brake van and two box vans acting as barrier wagons, No 92153 heads a train of oil tanks near Burscough on 11th October 1967. Following six months storage in a serviceable condition at Speke Junction early in 1966, the locomotive was re-activated to see out the last few months of its career – note the depot code 8C painted on the smokebox; the locomotive carries a class C head code. The locomotive would be withdrawn in January 1968 and consigned to South Wales for scrapping. *Arnold W. Battson*

Above: Most images at Carlisle Kingmoor were taken in the yard, here however No 92155 is seen inside the shed on 19th September 1966. Delivered from Crewe Works on 5th May 1957, it was allocated to Saltley, and transferred to Speke Junction in August 1966. No 92155 is one of a large number of 9Fs that is not recorded as ever having a General overhaul. Two months after the image was taken it would be withdrawn and, following storage at Speke Junction, would be dragged to Draper's yard at Hull for recycling. *James L. Stevenson*

Above: No 92156 is seen in the shed yard during a visit to Canklow, Rotherham, on 14th September 1958 whilst allocated to Toton, where it had arrived new from Crewe in November 1957. Canklow was home to numerous 'Austerity' 2-8-0s and Stanier 8Fs, however, no 9Fs were ever allocated there. The depot closed in June 1965. With the rundown of steam and closure of workshops, No 92156 is recorded as having a Light Casual overhaul at Eastleigh in October 1964 that helped see the locomotive through to withdrawal from Warrington in July 1967; it was dismantled at Draper's Hull yard on 18th March 1968.

Opposite Top: With a motley assortment of carriage stock in tow, No 92157 is at the head of the 1.50pm from Blackpool Central and approaches Kirkham & Wesham station on 28th June 1958, during the period it was allocated to Toton depot. To the west of the station, Kirkham North junction is where the suburban branch line to Blackpool South follows the Fylde coast through Lytham, Ansdell & Fairhaven and St Annes-on-the-Sea; the main line to Blackpool North proceeds via Poulton. Between 1903 and 1965 there was a third line, the 'Marton Line', which went straight to Blackpool South and beyond to Blackpool Central. This junction involved a flyover to allow Preston-bound trains to access the Up Fast line from the Marton line. Although the Marton line closed in 1965, the disused flyover bridge was not removed until the 1980s. Following its stint at Toton, No 92157 was moved via Saltley to Birkenhead where it arrived in April 1964, withdrawal came in August 1967 followed by scrapping in South Wales by J. Buttigieg in Newport. *Arnold W. Battson*

Opposite Bottom: No 92159 stands on shed at Cricklewood in July 1959 having worked in from its home depot of Wellingborough. It was delivered from Crewe Works on 27th November 1957 and spent over six years there until moved to Rowsley in February 1964 – although it is recorded as allocated to Cricklewood for a short period in December 1958. A final move to Birkenhead occurred in May 1965 with withdrawal in July 1967. T. J. Thompson of Stockton-on-Tees scrapped it in January 1968. *Jack A. C. Kirke*

Opposite Top: Wellingborough depot on 21st December 1957 with the recently delivered No 92160 raising steam alongside No 92019. To the right is the unmistakable blast pipe arrangement of a Crosti-fitted member of the class, in this case No 92026. No 92160 transferred to Kettering in September 1968. Withdrawn from Carnforth in late June 1968, it was sold to G. H. Campbell of Airdrie but scrapped at Gateshead by Clayton & Davie that October. *Ron F. Smith*

Opposite Bottom: Speke Junction depot was situated in the triangle of the Allerton to Garston Docks and Ditton junction lines to the west of Speke station. On 3rd July 1965 No 92161 stands in front of a rebuilt Crosti, with No 92008 alongside. No 92161 entered service at Westhouses on 17th December 1957, and, moving via Newton Heath, it ended its days at Carlisle Kingmoor in December 1966. The London & North Western Railway opened the 12-road shed on 10th May 1886. BR added a mechanical loading plant in the 1950s, with closure taking place on 6th May 1968; the depot was one of the storage points for withdrawn locomotives on their way to the scrapyards. *James L. Stevenson*

Above: With a 55ft 11in wheelbase the 9Fs were a tight fit on a 60ft turntable, as seen here at New Mills on 14th June 1961. This was a locomotive servicing point, with no depot facilities available, on the Great Central & Midland Joint Committee's line. The nearby New Mills station lies at the junction of what was the Hayfield branch and the Midland line; the two appearing through tunnels on a ledge blasted out of the cliff face, some 40ft above the River Goyt. The Manchester, Buxton, Matlock & Midlands Junction Railway ran as far as Rowsley and was extended by the Midland Railway to Buxton, in its aim to run as far as Manchester. The Manchester, Sheffield & Lincolnshire Railway (later Great Central) also wished to extend southwards from its main line through Woodhead Tunnel to Sheffield and had built a branch to Hyde. Meanwhile, the London & North Western Railway had constructed their own line to Buxton from Whaley Bridge, with a station at Newtown, which effectively blocked the other two. An agreement was reached whereby the MS&LR would build their proposed 'Marple, New Mills & Hayfield Junction Railway', while the Midland Railway would extend its line to New Mills from Miller's Dale via Chinley. Passenger services began to Hayfield in 1868 and the line came under joint control as the Sheffield & Midland Railway Companies' Committee in 1870, while the Midland's line opened in 1867. No 92162 emerged from Crewe Works in December 1957, being allocated to Westhouses; moving to Newton Heath in the summer of 1958, it was withdrawn from Birkenhead in early November 1967 and was taken to G. H. Campbell's Airdrie scrapyard for recycling in May 1968. *Alan H. Roscoe*

Opposite Top: About to pass under Wakefield Road overbridge, No 92165 is seen at Stourton during 1961 at the head of a class C service as it approaches the junction where the North Midland and East & West Yorkshire lines connected – the latter's route to Lofthouses closed to all traffic in 1966. New to traffic on 17th April 1958, No 92165 was sent to Saltley, where it was allocated at the time of the photograph, before moving to Bidston depot in June 1962 and ultimately being withdrawn from Speke Junction in August 1967. J. Cashmore of Newport scrapped it in June 1968. The photographer has his back to Stourton depot that was opened by the Midland Railway in 1893 and closed in January 1967; although home to a number of LMS heavy freight locomotives, no 9Fs were ever allocated there.

Opposite Bottom: An up freight passes Escrick on 27th September 1960 behind No 92168. The locomotive entered traffic from Crewe Works on 20th December 1957, being allocated to Doncaster, remaining there until withdrawn at the end of June 1965. Note the two wheel sets and framework in the foreground that can form a rudimentary trolley. The North Eastern Railway opened Escrick station on 2nd January 1871, and it was a victim of the pre-Beeching Report closures, being closed to passengers on 8th June 1953, and to goods traffic on 11th September 1961. The line was closed in October 1983, with all trains diverted onto the new section of the East Coast main line between Templehirst junction and Colton junction – the Selby diversion. *Alan H. Roscoe*

Above: As with a number of larger depots the cast concrete coaling tower – or cenotaph, as they were sometimes referred to – looms large over King's Cross's 'Top Shed' depot. For what it was like to run a major London steam shed readers are referred to Peter Townend's book on the depot – *Top Shed: Pictorial History of King's Cross Locomotive Depot*, Ian Allan Publishing, 1989, ISBN 9780711018273. No 92170 had worked in from its home depot of Doncaster, in July 1958, where it had arrived direct from Crewe Works in December 1957. It remained there until placed into store in January 1964 and, withdrawn the following May, went to T. W. Ward's scrapyard at Killamarsh for recycling. Alongside No 92170 stands one of Gresley's racehorses. *Jack A. C. Kirke*

Opposite: No 92178 leads a raft of light engines heading for the freight yards at Banbury consisting of another 9F and a GW tank locomotive – there was a hump marshalling yard here as it was a junction for former GW, LMS and Great Central lines so one could see locomotives working in over these routes. The station in the background is the GWR's Banbury station that gained the suffix General in 1949 to differentiate it from the LMS's Merton Street station that is hidden behind the locomotives. The latter station closed to passengers on 2nd January 1961 with services being diverted into General; freight traffic ceased on 6th June 1966 although services over the LMS route had closed in 1963. The GC line joined the GW line to the north of the town and passenger services used the GW station. Note the people walking along the track, no high-visibility clothing in those days.

Above: New to Peterborough New England in October 1957, No 92179 remained at the depot until moving to Langwith Junction in January 1965. It is seen here emerging from Hadley Wood North Tunnel at the head of a class C train consisting of express-passenger rated stock, the crew no doubt being pleased to reach fresh air. The station at Hadley Wood was sandwiched between the two tunnels – the north was 200yds (182.8m) in length with the south the longer at 384yds (351.1m). Withdrawn from Colwick depot on 14th November 1965 and scrapped by Hughes Bolckow of Blyth, it was one of only two 9Fs scrapped at the yard – the other was No 92221.

Above: A formal agreement to build a station at Harringay was made between the British Land Company and the Great Northern Railway in April 1884. The Land Company needed the station to serve housing it was building to the east of the railway line on the site of Harringay House, so it contributed £3,500 to the cost and agreed to bear the working costs of the station for an initial period. The station opened to passenger traffic on 1st May 1885. The station was renamed Harringay West on 18th June 1951, but reverted to Harringay on 27th May 1971. There were expansions of the track layout in 1888 when a new Down Goods line was brought into use, running behind the down platform on the formation of a siding; plus Up Goods Nos 1 and 2 lines from Hornsey, of which the former continued to Finsbury Park. These extra lines had been allowed for in the station design, but still required alterations to the goods yard connections and the removal of the existing signal box that was in the way. Part of the upgrade was the construction of replacement signal boxes – note the height of Harringay Up Goods 'box to enable the signalman to see over the adjacent footbridge giving access to the station from both sides of the line. Introduced to traffic in November 1957, No 92180 was withdrawn from Langwith Junction in April 1965.
David Stewart-David

Opposite Top: With a permanent way train in tow, No 92182 heads through Doncaster station with St James' Bridge in the background. The Great Northern Railway built the station in 1849, replacing a temporary structure constructed a year earlier. It was rebuilt in its present form in 1938 and has had several slight modifications since that date, most notably in 2006, when the new interchange and connection to the Frenchgate Centre opened. No 92182 entered traffic on 3rd December 1957, being sent to New England where it remained until January 1965 when it was transferred to Langwith Junction; it was withdrawn from Doncaster in April 1966. Note the water crane (behind the first wagon) enabling locomotives on the through road to take water. *Ian Strachan*

Opposite Bottom: At the head of a five carriage train made up of express-rated parcels stock, No 92183 is seen to the south of Rossington station with Middles Lane overbridge in the background on 18th August 1962. Rossington was on the line between Doncaster and Bawtry, then heading south via Retford, Newark, Grantham would take the locomotive home. No 92183 entered traffic on 20th December 1957 at New England before moving to Colwick in June 1963. It was withdrawn from Doncaster in April 1966 and as with No 92182 it was recycled by W. George of Station Steel, Wath, in July 1966.

Opposite Top: No 92163 is seen at the head of a southbound class H mixed freight train as it heads past Bedford South junction in June 1962 during its second stint at Kettering. It arrived at the depot in March 1958 following delivery from Crewe Works before moving to Leicester Midland for five months in June 1959, returning the following November. The original station at Bedford was built by the Midland Railway in 1859 on its line to the Great Northern at Hitchin. It was on land known as 'Freemen's Common' approximately 200yd (180m) south of the current station on Ashburnham Road. The London & North Western Railway also had a station on its line between Bletchley and Cambridge. The Midland crossed it on the level and there was a serious collision when an LNWR train passed a red signal. (Curiously, both drivers were named John Perkins.) Following this accident, the Midland built a flyover in 1885. No 92163 ended its days at the hands of G. H. Campbell cutters following withdrawal from Birkenhead in early November 1967. *Jack A. C. Kirke*

Opposite Bottom: No 92166 leads an up class E freight from South Wales across the River Severn at Over junction on 28th June 1965. Allocated new to Saltley, No 92166 spent a period on loan to Rugby Testing Station as it was one of three – Nos 92165-67 – to be fitted with Berkley mechanical stokers. It spent six months on loan to Ebbw Junction depot for trials on iron ore trains. Not wholly successful as it required coal of a specific grade, otherwise the fireman had to break the coal to reduce it to the necessary 5-6in (125-150mm) cube. The stokers were removed when the locomotives passed through Crewe Works in December 1962 and January 1963. Note the size of the lumps of coal on the tender, the fireman will be kept busy breaking them to fit through the fire hole door, the slacking pipe – used to dampen down the coal dust – is hanging out of the window. Behind the train is the junction of the Ledbury line, with the lines heading off the left hand side lead to Gloucester Docks that are located at the northern junction of the River Severn and the Gloucester & Sharpness Canal; they are Britain's most inland port. *Peter Kerslake*

Below: An up fitted class C freight is seen at Dent Head on 9th July 1961 behind Saltley allocated No 92167, that at the date of the photograph was still fitted with a Berkley mechanical stoker, enabling, on a good day, the fireman to have a slightly easier life. New from Crewe Works on 9th May 1958, it remained at Saltley, apart from a loan period to Tyne Dock as part of trials with the mechanical stoker, until transferred to Didston at the end of 1962. No 92167 had the distinction of being the last 9F in service, being withdrawn at the end of June 1968 from Carnforth depot; it was scrapped by Clayton & Davie's cutters in Gateshead that October. Nearby is Dent Head viaduct that is 596ft (182m) long, 100ft (30m) high, and consists of ten arches that are each 45ft (14m) across. The parapets of the viaduct are measured at 1,150ft (350m) above sea level. The spans are grouped into two sets of five, separated by a larger pier in the middle. The viaduct is constructed from Blue Limestone, which was quarried from Short Gill (quite near the viaduct itself) from the beds of Simonstone limestone, and another quarry almost underneath the viaduct. *Alan Robey*

Opposite Top: Holgate lies just to the south of York station on the former York & North Midland line. The platforms seen in the image are the remnants of York Racecourse station, also known as York Holgate Excursion Platform and Holgate Bridge station that served the York Racecourse in Holgate. The station was opened on 14th December 1860 by the North Eastern Railway, however, it was only used on race days; a cattle dock, no doubt for horse traffic, and a third platform were added on 25th September 1861. The sidings were used when the passenger numbers became too high, although this drew complaints due to delays and chaos. The station was last used for the races on 24th August 1939. No 92171 is seen heading south through the derelict station in charge of a permanent way train of rails, note the use of barrier wagons to protect the tender should the load shift during transit. The locomotive was new to Doncaster in February 1958, remaining there until transferred to New England in November 1963, from where, after a period in store, it was withdrawn at the end of May 1964.

Opposite Bottom: Having just traversed Welwyn South Tunnel, No 92172 wheels a rake of box vans through Welwyn North station. Construction of the station began in 1848 and the line was opened in 1850 as part of the Great Northern Railway. It was called Welwyn station until 17th July 1926, when it was renamed following the opening of a new station for Welwyn Garden City. It was built by contractor Thomas Brassey out of red brick produced locally from the Welwyn brickfields at Ayot Green. No 92172 left Crewe Works on 29th January 1958 bound for Doncaster, from where it was withdrawn in mid-April 1966. It was one of six members of the class to be scrapped by W. George of Station Steel at Wath-on-Dearne during June-August the same year.

Below: Working towards its own demise, No 92173 hauls a train load of concrete beams – sometimes these weighed more than 80 tons and were over 100ft in length – no doubt for bridge work during the construction of Britain's motorway network. The beam's transport involved the use of specially constructed wagons – in this case two six-wheel double-bogie vehicles separated by a two axle well wagon, all vehicles were close-coupled to ensure continuity of the draw gear (readers wanting more details are referred to *A Life with Locomotives*, D. W. Harvey, Marwood Publishing, 1992). No 92173 was new to traffic on 10th February 1958 being allocated Doncaster, moving to Langwith Junction in June 1965 then onto Colwick in October the same year. A final move in December 1965 saw it return to Doncaster from where it was withdrawn in March 1966 and scrapped at Ward's Killamarsh site the following May. The platforms seen in the image are the remnants of York Racecourse station mentioned opposite.

Opposite Top: No 92174 passes through York station on the avoiding lines during the summer of 1965 at the head of a southbound coal train. Introduced to traffic on 10th February 1958 from Crewe Works, it was allocated to Doncaster until withdrawal in December 1965. The first York railway station was a temporary wooden building on Queen Street outside the walls of the city, opened in 1839 by the York & North Midland Railway. It was succeeded in 1841, inside the walls, by what is now York old railway station. In due course, the requirement that through trains between London and Newcastle needed to reverse out of the old York station to continue their journey necessitated the construction of a new through station outside the walls. The present station, designed by the North Eastern Railway architects Thomas Prosser and William Peachey and built by Lucas Brothers, opened on 25th June 1877. It had 13 platforms and was at that time the largest in the world. As part of the new station project, the Royal Station Hotel (now The Principal York), designed by Peachey, opened in 1878. In 1909 new platforms were added, and in 1938 the current footbridge was built and the station resignalled. The building was heavily bombed during World War 2 and was extensively repaired in 1947. The station was designated as a Grade II* listed building in 1968. *Leslie Turner*

Opposite Bottom: Heck station was opened on 2nd January 1871 by the North Eastern Railway. It closed to passengers on 15th September 1958 and to goods on 29th April 1963, although sidings still served local companies producing building materials. The opening of RAF Snaith in 1941 increased the passenger traffic to and from the station. No 51 Squadron RAF arrived at the station in a special train, where the carriages were shunted into the sidings for unloading; it was reported that one of the carriages ran back onto the main line and caused a blockage. Today Heck is regretfully better known for an accident, on 28th February 2001, in which a vehicle with a trailer left the M62 carriageway and ran down the embankment, ending up obstructing the East Coast main line, resulting in the crash of a southbound Intercity service and a northbound freight train that left ten people dead. On 31st August 1962 No 92176 was caught heading south past the box with signalman Raymond Wardell looking on. New to Doncaster on 4th March 1958, the locomotive was withdrawn from Peterborough's New England depot at the end of May 1964 and sold to T. W. Ward, Killamarsh, for scrap. *Frank W. Smith (FWS472)*

Below: The first section of the Great Northern Railway (GNR) – that ran from Louth to a junction with the Manchester, Sheffield & Lincolnshire Railway at Grimsby – opened on 1st March 1848, but the southern section of the main line, between Maiden Lane and Peterborough, was not opened until August 1850. Potters Bar was one of the original stations, opening with the line on 7th August 1850. The station building, in a 'post modern' style, was the third on this site. In 1955 the station was described as 'The first of the Eastern Region's good modern stations'. The platform canopies were also constructed in 1955, using what was then an innovative technique of pre-stressed concrete. As the concrete set it unexpectedly curved up at either end of the long, thin canopies, unintentionally creating the 'willow' look. Heading a rake of express-passenger rated stock, No 92184 provides a contrast to the 'modern'-style station. The locomotive carries a 34E shed plate of Peterborough New England, dating the image to either June 1959-June 1963 or November 1963 to January 1964. It was withdrawn from Immingham in February 1964 and sold to Albert Draper of Hull for scrapping that took place on 10th June 1965.

Opposite Top: No 92185 is seen inside Swindon Works on 11th January 1958; it would enter traffic four days later, being allocated to New England depot at Peterborough. It moved to Colwick in June 1963 then back to New England that November before a move to Immingham in January 1964. It would be placed in store the following January before being condemned the following month. It was sold in May 1965 to Albert Drapers for breaking that took place within a few days of it arriving in Hull. *Ron F. Smith*

Opposite Bottom: No 92187 is seen light engine at Werrington junction during June 1963 (see page 67 for junction details). New from Swindon in February 1958, it was allocated to New England. It moved to Grantham in June, returning to the Peterborough depot that November. The locomotive has lost its shed code plate so the photo may have been taken around the time it was relocated to Colwick, that occurred during the month it was photographed. Withdrawn in February 1965, it was stored at Colwick before its final trip to Hull, being another of Draper's purchases. *Jack A. C. Kirke*

Below: An up train is seen at New Southgate, north London, behind No 92188 on 27th June 1960. Interestingly the Great Northern Railway station here was opened by order of the Middlesex Justices on 7th August 1850 as Colney Hatch & Southgate. The Justices insisted on trains stopping daily for the benefit of the Second Middlesex County Asylum – opened that year at Colney Hatch – that became known as Friern Hospital and was closed in 1993. At its height, the asylum was home to 2,500 mental patients and had the longest corridor in Britain (it would take a visitor more than two hours to walk the wards). For much of the 20th century, its name was synonymous among Londoners as it was used as a mental institution. No 92188 was delivered new to New England, and spent three months at Grantham (June-September 1958) before returning to stay until moved to Colwick in June 1963 from where it was withdrawn in February 1965 and consigned to Drapers of Hull for scrapping. *D. Idle*

Opposite Top: The original station at Grantham (Old Wharf) was opened when the Ambergate, Nottingham, Boston & Eastern Junction Railway opened its line from Nottingham on 15th July 1850; this was replaced by the present station that opened on 1st August 1852 – the Old Wharf station closed the following day. This line was taken over by the Great Northern Railway in 1854. No 92189 is seen passing through the station during the period it was allocated to Langwith Junction as it sports that depot's 40E shed plate. It had arrived there from Colwick in January 1965 only to return there the following October. Hidden by the lamp post is a poster advertising the Milk Marketing Board's promotion to 'drinka pinta milka day' that first appeared in 1959, remaining in use until the late 1970s.

Opposite Bottom: At the head of a rake of empty mineral wagons, No 92192 is seen just to the south of Woodford Halse on the Great Central Railway's London extension that opened here in 1898. Up until 1st November 1948 the station had been known as Woodford & Hinton. New to traffic from Swindon Works on 1st May 1958, No 92192 was sent to Doncaster where it remained until reallocated to Colwick in September 1963 for a couple of months to end its service life at Frodingham. Withdrawn in February 1965, by May it would be reduced to furnace-sized pieces by Ward's cutters to feed the steelworks. The overbridge is part of the East & West Junction Railway that opened in 1873, becoming part of the Stratford-Upon-Avon & Midland Junction Railway in 1909. Both routes succumbed to the massive closures of the 1960s. *P. Moseley (PM2136)*

Above: Seen on its home shed of Doncaster on 16th June 1958, No 92193 had arrived from Swindon Works the previous month. Transferred to Immingham in February the following year, it would remain there until withdrawn in June 1965 to become another of Draper's victims that November. No 92193 would under-go its one and only General overhaul at Darlington Works between 5th September and 6th October 1961. Note the open doors on the ash pans that enabled the fireman to empty them. *John Robertson*

Opposite Top: No 92196 was only two days into its short life when it was photographed near Newark on 20th August 1958 at the head of a mixed freight service. New to Doncaster from Swindon Works, its one and only transfer was to Immingham in January 1959. As with many 'late-build' locomotives it had only one major overhaul, in this a case a General repair at Darlington between 14th February and 8th April 1961. Withdrawn in early December 1964, it was stored at its home depot until consigned to Wards of Beighton for disposal. *John N. Smith (JNS646)*

Opposite Bottom: Hitchin was one of the original stations, opening with the line on 7th August 1850. On 21st October 1850, Hitchin became a junction station with the opening of the first section of the Royston & Hitchin Railway, between Hitchin and Royston (it was extended to Shepreth on 3rd August 1851). The Midland Railway opened a route from Leicester via Bedford to Hitchin on 1st February 1858, by which MR trains used the GNR to reach London. Immingham-allocated No 92197 is passing Hitchin station on the through lines at the head of a mixed freight on 20th April 1960. Delivered new from Swindon in September 1958, it was sent to Doncaster before moving to Frodingham for a couple of months in early 1959, then back to Doncaster before a final move to Immingham in September 1960. It was barely six years old on withdrawal in September 1965 and consigned to T. W. Ward at Beighton for scrapping. *Brian Wadey*

Above: Wearing a 36A shed code, No 92198 was allocated to Doncaster when it was new to traffic on 6th October 1958 and, apart from a short period at Frodingham early in 1959, remained there until sent to Colwick in September 1963. It is seen here on the through lines at Welwyn Garden City sometime during the period it was allocated to Doncaster, the station here was opened on 20th September 1926. This was the second garden city in England (founded in 1920) and one of the first new towns as designated in 1948. Following an all too short career, No 92198 was withdrawn early in August 1964 after a short period in store. The advertising hoardings were, and still are, a valuable source of revenue. Those above the platform trolley promote 'Imperial' (either a brand of soaps, toiletries and healthcare products, or cigarettes, the lettering is not clear), 'Sunblest – The sign of good bread, fresh to the last slice' along with 'Guiness for strength'.

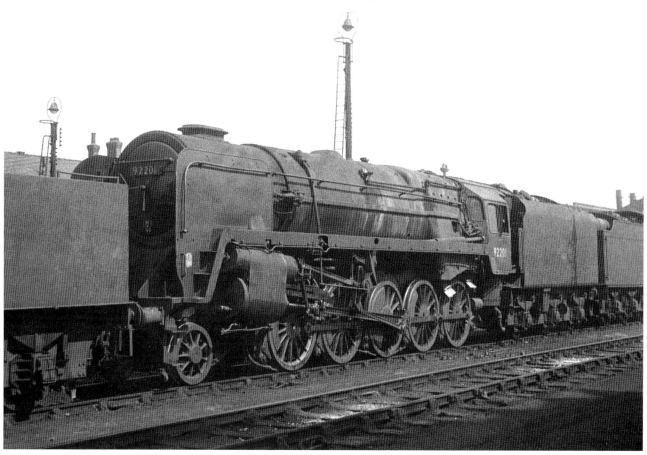

Opposite Top: No 92199 heads south through Dringhouses with a freight train on the approaches to Chaloners Whin junction. The junction was where the original line to Selby diverged from the York & North Midland line until its closure in 1983; see No 92040, page 22 for full details. Delivered new to Doncaster from Swindon in October 1958, it was reallocated to Colwick in September 1963, moving to Frodingham in November the same year. Withdrawal came in early August 1964 following a couple of months in store, it became another of the class to be scrapped by T. W. Ward at Beighton, Sheffield. Note the 'feathers' on the colour light signals which denote the route to be taken.

Opposite Bottom: No 92200 hauls a mixed freight train northwards past the box at Retford South, that is between Newark and Doncaster. The line off to the left connects this former Great Northern line to the Manchester, Sheffield & Lincolnshire (later Great Central) Railway's east-west route from Lincoln to Sheffield. The connection opened in 1849, surviving until closure in 1965. The train is interesting as the first vehicle is a 'Conflat' with the container covered with a tarpaulin, the fourth vehicle has a framework for carrying large flat sheets – usually metal, the sixth is a fuel tanker numbered 1692 on the tank end, then later in the rake three two-wheeled tank wagons are carried as part of the consist. No 92200 was delivered new to Doncaster in November 1958, and was withdrawn from Langwith Junction depot in October 1965, ending its days at T. W. Ward's Killamarsh yard.

Above: New from Swindon Works on 5th December 1958, No 92201 was sent to Doncaster where it is seen in a 'well used' condition during June 1962. In September 1963 it was transferred to Immingham, returning to Doncaster in June 1964. Withdrawn at the end of March 1966, it was sold to W. George of Station Steel, Wath, for disposal. Obvious in this view is the addition of conduit, fitted immediately below the running plate that carried the wiring for the automatic warning system – this was fitted only on the driver's side. *John F. Meakin (JFM237)*

Opposite Top: The driver of No 92222 has spotted the cameraman as the locomotive hauls a loaded iron ore train up Hatton Bank on 24th July 1958. The bank is five miles (8km) long with a ruling gradient of 1 in 110 for northbound trains, this section was often difficult to negotiate for heavy freights and the use of banking engines was commonplace – the photographer recorded the fact that this train was banked by GWR 2-6-2T No 5167 (withdrawn in January 1962). At the date of the photograph the 9F was less than two months old having arrived at Banbury depot, being new to traffic on 2nd June 1958. Withdrawal would occur at Southall depot on 26th March 1965, followed by a trip to South Wales for recycling. *Brian Wadey*

Opposite Bottom: With its withdrawal a matter of a few months away, No 92224 takes water at the north end of Preston station during June 1967. The image is full of detail from the railway infrastructure to the decorative panelling on the Fishergate bridge – generally this would have been unseen, but in this case it was viewable from the platforms. The locomotive was new to traffic from Crewe Works on 16th June 1958, being allocated to Banbury. Having been shuttled around various Western Region depots, it arrived at Warrington in August 1966 from where it was withdrawn at the end of September 1967. *Arnold W. Battson*

Above: Sporting a mix of GW end-on (on smokebox door) and the standard side-on lamp brackets (above buffer beam), Crewe-built No 92227 stands outside Crewe South depot on 18th August 1966 having worked north from Banbury shed where it had arrived for a second stay in August 1962. Following closure of Banbury depot in October 1966, the locomotive was moved to Warrington then onto Speke Junction from where it was withdrawn in November 1967. Following sale to T. W Ward the following January, it was scrapped virtually immediately on arrival at its Beighton yard. *James L. Stevenson*

Opposite Top: Brought to a halt by an adverse signal, No 92228 stands at Leamington Spa during 1959. New to traffic from Crewe Works on 12th July 1958, the locomotive was allocated to Banbury, remaining there until its one and only transfer to Speke Junction from where it was withdrawn at the end of February 1967. It was scrapped by T. W. Ward at Beighton, Sheffield the following June. The first station at the site, under the name Leamington, was opened by the Great Western Railway on its new main line between Birmingham, Oxford and London in 1852. It was later renamed Leamington Spa in 1913. This was not the first station in Leamington as the London & North Western Railway had reached the town eight years earlier in 1844, with a branch line from Coventry. That line, however, terminated about 0.7 miles (1.1km) from the town centre, at Milverton station. The opening of the GWR line compelled the LNWR to extend their Coventry branch into the centre of Leamington, and join it end-on to their new branch to Rugby, and in 1854 they opened a new station directly alongside the GWR station known as Leamington Spa (Avenue). In 1864, a connection was made between the GWR and LNWR lines at Leamington, which was mainly used to exchange goods traffic. The GWR station was often referred to as Leamington Spa General in order to distinguish it from the adjacent LNWR station, however it only carried this name officially between 1950 and 1968.

Opposite Bottom: No 92230 is seen on shed at Worcester on 26th July 1964 whilst allocated to Bromsgrove. It had arrived at Banbury, new from Crewe Works, in August 1958. It spent two periods allocated to Old Oak Common and three stints at Newport Ebbw Junction. It was withdrawn from Gloucester Horton Road on 31st December 1965 at the end of steam on the Western Region; it returned to South Wales for recycling. It is standing in front of an unidentified Western Region 'County' class 4-6-0, although as only one example was still in traffic at this date it may well be No 1011 County of Chester. David Birt

Above: Surrounded by the new order, No 92231 is seen on shed at York on the 18th April 1964 sporting a mixture of GW side on irons (on the smokebox door) and standard (side-on) lamp brackets. Delivered new to Pontypool Road in August 1958, it moved around the WR, with stays at Severn Tunnel Junction, Cardiff Canton, Newport Ebbw Junction and Bristol Barrow Road. In December 1960, it was then loaned to the Southern Region, before being allocated to Eastleigh in January 1961, moving on to Feltham in May 1963. Its move to York came in September 1963, remaining there until withdrawn in the middle of November 1966. After a period in store at York it went to Albert Draper in Hull for recycling that took place on 24th April 1967. J. T. Clewley

Above: At the head of a northbound freight consisting of empty mineral wagons, No 92233 is seen near Lamington on 17th July 1965. The line was constructed by the Caledonian Railway, opening on 15th February 1848; it follows the route of the River Clyde from south of Elvanfoot to north of Lamerton. It remains open and now sees Pendolinos running on the electrified route. No 92233 left Crewe Works on 11th August 1958 and was consigned to the Western Region at Pontypool Road. It headed north in June 1964 when it was sent to Carlisle Kingmoor. It was transferred to Speke Junction during January 1968 and withdrawn virtually immediately, being sold and scrapped that August. *James L. Stevenson*

Opposite Top: The Stephenson Locomotive Society was busy on 5th March 1967 when it ran two specials from Tyseley to Birmingham Snow Hill. The first train used No 7029 *Clun Castle* from Tyseley to Birkenhead, with No 44680 working the return, the second train used the same locomotives but with No 44680 out and 7029 return. The first train used No 92234 for a round trip from Birkenhead to Chester General and back via Hooton, the photographer did not record the actual location, but No 92234 has been cleaned and the smokebox fittings and buffer heads have been specially painted for the occasion. The second train used No 92203 on the same route. *R. F. Smith*

Opposite Bottom: Delivered new from Crewe Works to the Western Region in May 1958, its original allocation was Pontypool Road, remaining there until the end of the year when it was transferred to Severn Tunnel Junction. It is recorded as undergoing a Heavy Intermediate overhaul at Swindon between 4th May and 17th September 1962, dating the image towards the end of that time frame. Withdrawn from Bristol Barrow Road on 18th November 1965, No 92235 was stored at Oxford before making its final journey back to South Wales for breaking up. *S. J. Cowley*

Above: No 92237 received a Heavy Intermediate overhaul at Swindon Works between 4th January and 8th May 1962. Work had been completed by the 6th May as it is seen here awaiting its return to traffic a couple of days later. It had been completed at Crewe Works, entering traffic on 9th September 1958, to be withdrawn seven years later on the 10th September 1965. It arrived at Newport Ebbw Junction depot in September 1958, moving on to Cardiff Canton, Cardiff East Dock and Severn Tunnel Junction before a return to Newport in February 1965. Surviving in traffic until that September, it met its fate in John Cashmore's yard at Newport two months later.

Opposite Top: The crew of No 92238 take water at Evercreech Junction during a Warwickshire Railway Society rail tour on 12th June 1965. The 'Somerset & Dorset Joint & Eastleigh Tour' commenced at Birmingham Snow Hill behind Stanier Class 5 No 44777, with No 92238 running from Bath Green Park to Bournemouth Central where No 34097 *Holsworthy* took the train onto Eastleigh. After the visit No 34097 took the train on to Oxford with No 6967 *Willesley Hall* running back to Birmingham. This was the final passenger working of a 9F over the S&D, as No 92238 returned light engine to Bath. No 6967 would be withdrawn in December 1965; the 9F, however, had already been withdrawn from Severn Tunnel Junction the previous September, which probably is not surprising as it is recorded that it was not in the best condition during its climb over the Mendip hills.

Opposite Bottom: No 92241 stands outside Crewe South shed having worked in from South Wales. The shed was opened by the LNWR on 1st October 1897, being located to the west side of the Stafford line to the south of the station. It was closed on 6th November 1967. No 92241 was delivered new to Newport Ebbw Junction in October, with a move to Old Oak Common shortly after. Judging by the state of the locomotive, particularly the somewhat bent rails on the smoke deflectors, it was probably taken during the period it was allocated to Cardiff Canton, from November 1960 or Cardiff East Dock where it arrived in September 1962 following the closure of Canton to steam. It left the latter in November 1963 bound for Southall where it was withdrawn on 2nd July 1965. It returned to South Wales for scrapping, being cut up in Newport by J. Cashmore that December. Alongside stands Stanier Class 5MT No 45202, built by Armstrong Whitworth in October 1935, that remained in traffic until June 1968.

Above: No 92242 is seen on former GW metals as it heads southbound out of Gloucester on 1st June 1963. The two tracks nearest the cameraman were the GWR's, with the next two forming the Midland Railway's. The fifth track, running southbound, was the Midland Railway's New Dock branch that opened in 1900 serving Hempstead Wharf and Docks Branch West; closure of the final section of this branch took place in 1971. Going north the MR line served Gloucester station from opening on 12th April 1896 (Eastgate from 17th September 1951) until closure on 1st December 1975. No 92242 was allocated to Newport Ebbw Junction on delivery from Crewe in October 1958; moving to Severn Tunnel Junction in October 1964, it survived in traffic until May 1965. The consist behind No 92242 is remarkably tidy with the first nine wagons containing brick, the mineral wagons probably coal with tank wagons bringing up the rear. *Brian Wadey*

Opposite Top: New to traffic from Crewe Works on 28th October 1958, No 92244 was allocated to Newport Ebbw Junction, moving east to London's Old Oak Common the following month. It returned to South Wales in November 1960 when it arrived at Cardiff Canton, and, apart from a short stint at Oxford in the autumn of 1962, it spent the following five years allocated to Welsh depots. Its last recorded overhaul was a Heavy Intermediate one that took place at Swindon between 20th February and 23rd May 1962; it is seen at Llanelly a little over two years later on 26th July 1964 badly in need of a clean. Returning to England in October 1965, it was withdrawn from Gloucester Horton Road at the end of the year on 31st December 1965. It returned to South Wales for scrapping the following year. By comparison the English Electric Type 3, later Class 37, No 6858 was introduced in August 1963 and withdrawn in September 2007 – a service life of 44 years, a lot better than the seven years of the 9F. *Leslie Turner*

Opposite Bottom: With the passengers anticipating a few days by the sea, the southbound 'Pines Express' departs Broadstone behind No 92245 on the 8th August 1962. This was the only year the locomotive worked the summer season on the S&D having arrived at Bath Green Park in June 1962, relocating to Oxford in October. The 9Fs first arrived on the S&D for the summer season in 1960 as the heavy trains and topography of the route dictated most long distance services needing to be double-headed – the 9Fs being more than able to handle the trains by themselves. If a single-chimney locomotive was seen on the S&D it had clearly been borrowed as only those with double chimneys were ever allocated to Bath Green Park. *Brian Wadey*

Top: No 92246 is at the head of a class F freight train near Reading in September 1962; both tracks have been re-laid with concrete sleepers and flat bottom rail. The contents of the flat wagons at the end of the rake can't be clearly identified but may well be farm implements destined for a dealer. No 92246 emerged from Crewe Works, entering traffic on 20th November 1958, being allocated to Old Oak Common, moving around various Western Region depots over the next seven years. It underwent its only Heavy Intermediate overhaul at Swindon from April to July 1962, although it underwent numerous Unclassified repairs at various depots. It was withdrawn on 31st December 1965 from Gloucester Horton Road and scrapped in South Wales the following April. *Alan H. Roscoe*

Bottom: No 92247 stands head to head with No 92004 on Banbury shed in June 1966 having had its ash pans raked out. No 92247 arrived on the Western Region in December 1958 following its delivery from Crewe Works, being allocated to Old Oak Common, remaining there until February 1962 when it was moved to Cardiff Canton. It arrived at Banbury shed in August 1962, remaining there until a final move to Newton Heath in September 1966. Following its withdrawal in early October 1966, No 92247 was scrapped by Albert Draper's cutters at Hull on 10th April 1967. *Jack A. C. Kirke*

Left: Having conquered Beattock summit, No 92249 runs into Carstairs station at the head of a rake of hoppers. The original station here was opened by the Caledonian Railway on 15th February when the line between Glasgow and Beattock was completed. The line from Edinburgh reached Carstairs and opened on 1st April 1848 and this can be seen disappearing off to the left behind Carstairs No 3 box. Between 1914 and 1916 the Caledonian Railway began an extensive reconstruction of the station. The original layout was constricted with only one set of up and down through lines and was expanded with the provision of loop lines for both up and down trains to allow non-stop trains to bypass any trains stopped at the station. Interestingly from 1888 to 1895 the station was also the terminus of the Carstairs House Tramway that connected to Carstairs House. Delivered new to Newport Ebbw Junction, No 92249 remained on the Western Region until October 1963 when it was transferred to Newton Heath. At the time of the photograph it carries a Carlisle Kingmoor shed plate having arrived there in June 1964. Its final transfer to Speke Junction depot occurred in January 1968 from where it was withdrawn in May 1968 – one of the last to remain in service. *Ian Strachan*

Below: Numerically the last of the class actually entered service from Crewe Works in December 1958, being allocated to Banbury. It entered traffic sporting a Giesl Oblong Ejector in place of the by now standard double chimney. This was the last significant development in chimney/blast pipe configuration and was devised by Dr Adolph Giesl-Gieslingen of the Vienna Technical University in collaboration with Austrian State Railways. Compared with results in Austria, trials at the Rugby Testing Station showed little improvement over the standard design. Perhaps by the time the later 9Fs entered service, BR steam had reached the pinnacle of design. No 92250 was reallocated to Southall depot in November 1963, where it is seen on the 8th December, returning to Newport Ebbw Junction the following May. A final relocation to Gloucester Horton Road occurred in October 1965 from where it was withdrawn on the 31st December that year. It retained the Giesl ejector until the end that occurred at the hands of John Cashmore's cutters at Newport in July 1966. *Brian Wadey*

Above: No 92203 emerged from Swindon Works as the first of the class to be built under Lot 429, entering traffic at Bristol St. Philip's Marsh on 6th April 1959. It is seen here south of Farington, the LMS one – not Farrington on the GWR – with an up freight train on 2nd October 1967. The station is just beyond the overbridge in the background and was opened by the North Union Railway on 31st October 1838 – albeit spelt Farrington – closing to all traffic on 7th March 1960 before Dr Beeching wielded his axe. The area was a railway crossroads to the south of Preston, with lines to the north, south, east and west, it was also home to Lostock Hall locomotive depot. Following its withdrawal from Birkenhead depot in early November 1967, No 92203 was sold to the well-known artist David Shepherd and can now be found on the North Norfolk Railway. *Arnold W. Battson*

Below: The Saturday's only Bradford-Bournemouth service is hauled through Wincanton by No 92204 on the 30th July 1960 – the first summer of 9Fs on the Somerset & Dorset. No 92204 entered traffic from Swindon Works on 21st April 1959, being allocated to Bristol St Philip's, transferring to Bath Green Park in June 1960 and is seen here sporting its 82F shed plate. Wincanton station was situated on a double-track section of the S&D, opening in November 1861 as part of the Dorset Central Railway. Goods traffic ceased on 5th April 1965 with final closure to all traffic on 7th March 1966 when the Western Region closed the line. No 92204 was withdrawn from Speke Junction depot in early December 1967. *Brian Wadey*

Above: No 92205 is seen at the head of a Bournemouth-bound service at Templecombe junction in July 1960. The locomotive would haul the train into Templecombe's main station on the LSWR's main line, then, with a pilot engine, reverse back to the junction to continue its journey southwards. It would pass through Templecombe's Lower station that was located on the S&D Joint line; opening on 3rd February 1862 it would survive until 3rd January 1966, with the South Western station closing on 7th March 1966 – along with the S&D line. The delightfully named Combe Throop Lane connected the two stations. The station on the former LSWR main line was reopened on 3rd October 1983 following community pressure. No 92205 entered traffic on 1st May 1959 and was one of five locomotives to be allocated to Eastleigh depot in 1961. It was transferred to its final depot of Wakefield in October 1966, and following a period in store there was withdrawn in early June 1967 and consigned to Drapers of Hull for disposal. *Jack A. C. Kirke*

Below: A 350hp shunter (later Class 08) tries to hide behind No 92208 whilst it shunts the goods yard to the west of Preston station on 5th May 1967. Interestingly the first rail lines in Preston were those of the Lancaster Canal Tramroad, a horse-drawn line connecting two parts of the Lancaster Canal. It opened in 1805, but never carried passengers or converted to steam; it ceased operating in Preston in 1862. The first steam-hauled passenger railway in Preston was the North Union Railway. On 31st October 1838, it opened its line from Wigan to a station on the site of the present-day Preston Station. This

immediately linked the town to London, Birmingham, Liverpool and Manchester. A different company built each subsequent line and rivalry prevented any cooperation over shared facilities, and so almost every railway line into Preston used its own station. It was not until 1900 that all lines in Preston shared a single station, by which time all the railways had been amalgamated into just two companies. No 92208's first depot was Laira, Plymouth, where it arrived from Swindon Works in early June 1959. It stayed on the Western Region until October 1963 when it was transferred to Newton Heath. Withdrawn from Carlisle Kingmoor at the end of November 1967, it was scrapped in February 1968 by J. McWilliam in Shettleston.
Arnold W. Battson

Opposite Top: No 92209 hauls the Plymouth-Crewe service out of Teignmouth and along the South Devon coast wall in June 1959. The station was opened by the South Devon Railway (SDR) on 30th May 1846 as the terminus of its first section from Exeter. The line was extended to Newton Abbot on 31st December 1846. Teignmouth's single platform was augmented by a second one late in 1848 – at this time it was a 7ft 0.25in (2,134mm) broad gauge railway. Teignmouth was the original headquarters of the SDR, the station and offices being described as a 'primitive apology for a station' and locally dubbed the 'Noah's Ark'. On 20th May 1892, the line was converted to 4ft 8.5in (1,435mm) standard gauge. The station was completely rebuilt soon after, the work being completed early in 1895. At the date of the image No 92209 was virtually new having arrived at Laira, Plymouth, a few days earlier; it remained allocated to the Western Region for the rest of its career, being withdrawn from Bath Green Park in December 1965; although this may have been a 'paper transfer' from Bristol St Philip's depot where it was stored before ending its days at the Newport yard of J. Cashmores. The photographer was lucky as the rear of a down service is disappearing into the distance. *Stanley Creer*

Opposite Bottom: At the head of a down oil tank train No 92212 passes through Basingstoke at 6.46pm on the 8th September 1962, note the barrier (or buffer) wagons between locomotive tender and leading tanker. These oil trains were worked for Esso Petroleum who had a refinery at Fawley, on the Solent, to Bromford Bridge distribution centre near Birmingham, usually running via the former Didcot, Newbury & Southampton line until that route closed. Five 9Fs were allocated to Eastleigh for the services, Nos 92205/6, 92111, 31 and 39, so with Tyseley-allocated No 92212 in charge, a failure en-route may have occurred. The Southern 9Fs were allocated away from the region in September 1963, although 9Fs continued to appear on the Southern Region when there was a locomotive failure or for overhaul at Eastleigh Works. Withdrawn in early January 1968, No 92212 was preserved via Woodham's scrapyard in 1979 and is today owned by Locomotive Services Ltd. *David B. Clark*

Above: No doubt heading for the docks, No 92213 wheels a freight train made up from XP-rated box vans on the approaches to Basingstoke on 12th September 1964, with the signalman waving out of the 'box window. The station was opened by the London & South Western Railway as a temporary terminus when its line to Southampton reached Basingstoke from London. It became a through station when the section running north from Southampton was completed later in 1840. The intention to build a line from near Basingstoke to Bristol was dropped when the Great Western Railway route was approved. The L&SWR did, however, plan a line to Salisbury from Basingstoke but this was delayed by financial difficulties. Eventually, it was built, reaching Andover in 1854 and Salisbury three years later, before being extended to become the West of England Main Line. The Great Western Railway opened its broad gauge line from Reading on 1st November 1848 with a separate station north of the L&SWR station. After its conversion to mixed gauge, on 22nd December 1856, through services could run between Southampton and Reading. The broad gauge rail was removed on 1st April 1869. The GWR station was closed on 1st January 1932, since when trains from Reading have used the main station. No 92213 was taken into traffic on 22nd October 1959 at Bristol St Philip's Marsh; almost immediately transferring to Banbury, it was withdrawn from Carlisle Kingmoor in early November 1966. *Brian Wadey*

Opposite top: A 9F sandwich outside Swindon Works on 20th December 1959. No 92218 is between No 5932 *Haydon Hall* and an unidentified member of the class. It entered traffic on 18th January 1960, allocated to Bristol St Philip's Marsh where it would remain until September 1960 when it was transferred to Old Oak Common. Its final move was to Speke Junction depot in January 1968, and it was withdrawn in May 1968, being scrapped two months later by Arnott Young at Parkgate. *D. Idle*

Opposite bottom: Nos 92219 and 92220 inside Swindon Works on 20th December 1959 – as well as identification marks, No 92219 has 'Do not swipe merry xmas' on the buffer beam and No 92220's has 'the last one thank the Lord' and 'equality for the worker'. The smokebox saddle of No 92220 has 'mano the loco' and a scholar of Latin has chalked Filius Nuilius (Son of None) – obviously a die-hard Great Western man. Perhaps the Latin student should have written Ecclesia erat recta (Churchward was right)! No 92219 was ex-works on 27th January 1960 with No 92220 departing on 25th March 1960 following its naming ceremony that had taken place on the 18th. *D. Idle*

Below: The 20th September 1964 saw No 92220 *Evening Star* working the Stephenson Locomotive Society's 'Farewell to Steam' tour running an out-and-back trip from London Victoria. No 92220 took the train from Victoria to Yeovil Junction via the MidHants route and Southampton, where Ivatt 2-6-2T Nos 41206 and 41308 took over running to Yeovil Town then down to Seaton Junction for a trip over the branch. The 9F returned from Seaton Junction to London via Chertsey, Staines Central and Twickenham. *Evening Star* was withdrawn from Cardiff East Dock following collision damage, with a buffer stop, on 26th March 1965 after five years in traffic. There were several abortive attempts at private preservation, fortunately No 92220 survived long enough to be included under the Transport Act of 1968 as part of 'a collection maintained or proposed to be maintained by the Secretary of State'. It is now part of the National Collection, and is currently at the National Railway Museum, York. *J. G. Walmsley*

Out of Steam

Looking as if it could be ready for a return to service, No 92043 stands in the yard at Carlisle Kingmoor on 19th September 1966; it had however been withdrawn at the end of July. It entered traffic on 18th January 1955, being allocated to March depot then moving to Annesley in February 1957, before a final transfer to Carlisle Kingmoor in January 1966. October 1966 saw its sale to G. H. Campbell of Airdrie for scrap, where it was almost immediately reduced to furnace sized pieces. *James L. Stevenson*

Very much out of steam, having been stripped for its journey to the scrapyard, No 92072 stands forlorn at Birkenhead on 4th April 1966. It had been withdrawn from Kirkby-in-Ashfield depot in early January, and stored there until dragged to Birkenhead. It was sold in June 1966 and broken up later the same month by T. W. Ward at Beighton, Sheffield. Behind stands an unidentified Fairburn 2-6-4T with the mineral wagons probably containing the removed motion from the locomotives. Note that the 9F is missing a smoke deflector and has a shunter's pole pushed through the lifting eyes above the buffer beam. *James L. Stevenson*